The Human Challenge of Telemedicine

**Health Industrialization Set**

coordinated by
Bruno Salgues

# The Human Challenge of Telemedicine

*Toward Time-sensitive and Person-centered Ethics in Home Telecare*

Philippe Bardy

ISTE PRESS

ELSEVIER

First published 2019 in Great Britain and the United States by ISTE Press Ltd and Elsevier Ltd



ISTE Press Ltd
27-37 St George's Road
London SW19 4EU
UK

www.iste.co.uk

Elsevier Ltd
The Boulevard, Langford Lane
Kidlington, Oxford, OX5 1GB
UK

www.elsevier.com



British Library Cataloguing-in-Publication Data
A CIP record for this book is available from the British Library
Library of Congress Cataloging in Publication Data
A catalog record for this book is available from the Library of Congress
ISBN 978-1-78548-304-2

# Contents

# Acknowledgments

I would like to express my gratitude to all those who actively supported me at the various stages of writing this book: a revised and expanded version of my PhD thesis in philosophy of sciences defended at the University Paris 1 Panthéon-Sorbonne on December 8, 2016. I would also like to thank:

– Bernadette Bensaude-Vincent, my thesis supervisor, for her unfailing support;

– Xavier Guchet, Gérard Dubey and Bernard Reber, respectively thesis rapporteurs and president of the jury, for their enlightening analyses inspired by the philosophy and sociology of new technologies, applied ethics and moral philosophy;

– Bruno Salgues for his fine and meticulous proofreading of the original manuscript and for his valuable editorial recommendations;

– my comrades of the doctoral seminar organized by Bernadette Bensaude-Vincent (Jean Hache, Vincent Devictor, Gabriel Dorthe, Élise Aurières, Dorine Montout, Quentin Hardy, Jeanne Guien, Pierre de Jouvancourt) as well as members of the *Centre d'études des techniques, des connaissances et des pratiques* (Cetcopra) of the University Paris 1 Panthéon-Sorbonne; in particular Thierry Pillon, Sophie Poirot-Delpech, Jean-Paul Karsenty, Sacha Loeve, Baptiste Monsaingeon and Marina Maestrutti.

Many partners have also given me valuable support in my research and investigative work. All of them are warmly thanked here:

– Anne Pina, president of the association *Les Porteurs*, and Laurent Gaillard, general delegate, helped me, more than I could have hoped, to discover and share the daily reality of pacemaker and defibrillator carriers;

– Alain Perrin and Daniel Jouen, President and Vice-President of APODEC, greatly facilitated meeting with the members of their association;

– Agnès Caillette-Beaudoin and Jean-Pierre Grangier, respectively nephrologist and physician director at the Calydial Dialysis Unit based in Irigny and Calydial's health executive, as well as Hafedh Fessi, nephrologist at Tenon Hospital, and Professor Éric Renard, endocrinologist and diabetologist at Montpellier University Hospital Center, who actively supported me in my search for testimonials from patients using at-home dialysis and patients under autonomous insulin therapy.

# Introduction

## Temporalities and Psycho-Emotional Dimension of the Technicality of Care

*It has been the glory of modern science that it has been able to emancipate itself completely from all such anthropocentric, that is, truly humanistic, concerns.*

Hannah Arendt [ARE 07]

Based on an empirical study coupled with epistemological and ethical reflection on contemporary telemedicine, this book mainly questions the temporal regime and the psycho-emotional dimension of patients with chronic diseases who participate in remote monitoring experiments of their conditions in autonomous home management (renal and cardiac failure, diabetes). This unique practice of remote-assisted care, because of electronic prostheses and/or connected devices, is called telecare. Field data collected between 2012 and 2015, in France and abroad, from telemedicine stakeholders and people interviewed on this subject, show that telecare makes it possible to monitor patients' illnesses and provide care remotely to improve their overall quality of life, while still paying insufficient attention, in the general opinion, to patients' physical and psychological experiences. Similarly, limited attention is given in recent studies to the perception of caregiver time (time of care, conversation) or the temporal regime of patient interactions with telecare objects (care devices or communicating objects such as sensors, smartphones, tablets, etc.).

The contextualized analysis of these interactions, and more broadly of telecare protocols, nevertheless shows that the management of the multiple temporalities of the participants, and in particular those of the patient, can increase the scope of care and improve the support of the person on the disease pathway. In addition, an examination of current telecare practices reveals the (therapeutic) interest of a broader management of the psycho-emotional impacts of the disease, both positive and negative, and of certain factors separate from the disease, which also constitute the person's experience and the potential determination of their future: social and family environment, caregiver empathy, physical activities, etc.

Patient commitment, i.e. their willingness and capacity to become involved in remote medical monitoring protocols, is also based on the production and negotiated circulation in care receiver–caregiver interactions of moral feelings (hope, trust, respect) and specific care practices that constitute a "moral economy" of telecare, in the sense given to this expression by the historian of science Lorraine Daston, in other words: "A web of affect-saturated values that stand and function in well-defined relationship to one another". The American historian adds: "A moral economy is a balanced system of emotional forces, with equilibrium points and constraints" [DAS 95, p. 4]. Anthropologist Didier Fassin explains the meaning that Daston gives to the moral economy at the heart of scientific activity, transposable here to telemedical activity perceived from the point of view of health professionals:

> *Daston intends to look for the "affect-saturated values" that constitute the moral economies at the heart of scientific activity, in what she calls "Merton's black box", but we could just as well add the black boxes of Pierre Bourdieu, Jürgen Habermas, Michel Foucault, and a few others. It is neither the inwardness of scientists – their psychology and motivations – nor the exterior of science – ideologies, interests, powers – nor even a kind of genius of its own – ideal and immutable norms – that must be understood, but the values and affects that underlie the daily work of researchers, both individually and collectively, and that are likely to evolve over time and to differ from one society to another* [FAS 09].

The process of ethical standardization of telecare observed in the field, specific to this moral economy, is based on a set of care techniques that aim to empower the patient and make the patient accountable. However, this dual process of empowerment and responsibility is part of a temporality imposed, at least in part, by telecare (*timing* of self-measurement, compliance schedule, daily care programming). The phenomenological analysis of telecare technologies shows that autonomy (under monitoring), which is sought for the patient, aims to compensate for the *relational gap* that inevitably widens, under the effect of remote care, between the caregiver and the care receiver. However, this compensation should not be to the detriment of empathic patient care, as the testimonies collected from patients with chronic diseases show. In other words, *autonomy* and *relationship* cannot be mutually exclusive in a telecare context. The meaning of "caring" is based, quite to the contrary, on the balance that participants manage to find between these two arenas of telecare.

To frame this *work* of care, we propose to evaluate the contributions and limits of the ethics of autonomy, i.e. of ethics "based on a conception of the human being considered as a consciousness in full possession of their power, capable of expressing their freedom, their plan, their rationality and their desires and to impose them within an egalitarian contract" [TAN 13]. While autonomy is about patient empowerment, greater attention must be paid to the risks inherent in excessive delegation of responsibility for personal care and in an exacerbated requirement for compliance with protocol, which may oppose the patient's rationality and wishes. The evaluation of the limits of autonomous care practices makes it possible to envisage the relational ethics of "refocused" care on vulnerability and emotional health *as well as* the increased technical *and* human care of the *person.*

The hypothesis explored in this book is that a relational ethics of telemonitoring, without breaking with the values of autonomy and responsibility, makes it possible to define the conditions for a new therapeutic alliance that is respectful of people in their psycho-emotional dimension, their temporal regimes, their own values and their preferences. Based on a contextualized analysis of patients' daily practices and experiences, this book is a critical contribution to ethical reflection on telecare as a human practice that treats the individual as a "subject" of responsible care, respected in his or her values.

Telecare, as a "good practice" redefining the traditional care receiver–caregiver relationship, places at the center of contemporary care ethics *anthropocentric and truly humanisitic concerns* that, according to Hannah Arendt, have been *abandoned* by science and technology. These concerns, in the context of remote care, take many forms – concern, empathy, humanity of the medical team – which will be approached here from a triple phenomenological, epistemological and ethical angle.

The first part of this book is devoted to a contemporary inventory of digital health services in France in which telecare is accessed by patients suffering from chronic diseases. The specificities of telecare are important to identify, as it is being expanded with great support from legal texts, medico-economic evaluations and experience feedback. Based on three case studies, this phenomenological examination makes it possible to identify as closely as possible the ethical stakes of telecare as well as the values shared by the participants, as understood here from the prism of medical ethics and *care* ethics.

The "patient experience" analysis within telecare protocols, conducted in the second part, is based on the phenomenology of Maurice Merleau-Ponty's perception and Gilbert Simondon's philosophical psychology. It is based on a preliminary reexamination of the values of *temporality* and *sensitivity* belonging to the person involved in a protocol of care, which in turn sheds light on the phenomenological and epistemological meaning and scope of the "patient experience". This experience is notably that of a singular relationship between the patient and the multiple objects of telecare (care devices *and* communication devices) that redefines their relationship to the disease and the healthcare team. The testimonies collected allow us to assess the impacts of telecare, both positive and negative, on the remote care receiver and to explore the temporal and psycho-emotional dimensions of this experience.

The communication devices used in daily life allow the patient to anticipate the complications of their health condition and to adapt the efforts or reorganize their activities: these devices are thus, as we will see in the third part, *time-management devices*, put at the service of the person. At the same time, the management of care, and of its temporality, by interposed objects, cannot today be achieved without ensuring an emotional and relational management of remote care. The examination of the temporal and psycho-emotional dimensions of telecare linked to the use of prostheses and connected devices makes it possible precisely to envisage a "relational" and "chrono-sensitive" telecare ethics, i.e. ethics centred on practices concerned with the temporal *and* psycho-emotional system of people, within the framework of a remote care *relationship*. The aim of this ethics can be simply summarized thus: to guide and frame current telemedicine practices with the goal of reconciling humanity and the technicality of care.

# PART 1

# The Person in the Age of Telecare

# 1

# The Advent of Digital Healthcare

*The more the presence of medicine is based on technique, the more the subject to which it is addressed is fiction.*

Didier Sicard [SIC 02, p. 17]

## 1.1. Digital healthcare as an ecosystem

Telemedicine, i.e. the use of digital technologies by healthcare professionals for remote medical practice, has enjoyed national deployment since 2012, following its legal recognition in the *Hôpital, Patients, Santé, Territoires* (HPST: Hospital, Patients, Health, Territories) law of 2009 and the description of its regulatory framework in the decree of October 19, 2010 (see Box 1.1). Telemedicine does not replace current medical practices but is a response to the challenges facing healthcare provision today: inequalities in access to care, medical demography, decompartmentalization of the system and economic constraints. In France, professionals in the health and medico-social fields are developing an awareness of the need for a multidisciplinary, transversal and coordinated approach to improve the health pathways.[1]

The HPST Law – article 78 – amending article L.6316-1 of the French Public Health Code defines telemedicine as follows: "Telemedicine is a form of remote medical practice using information and communication technologies for screening and diagnosis, consultation, expertise, conference, monitoring, transmission of images and data, and therapeutic intervention". The law specifies that it puts in contact, with each other between them or with a patient, one or several health professionals among whom are necessarily a medical

1 See, for example, the summary of the Regional Telemedicine Program (RTP) of the ARS Île-de-France. 1. Diagnostic 2012, p. 5.

professional and, if necessary, other professionals bringing their care to the patient. Telemedicine makes it possible to establish a diagnosis, provide preventive or posttherapeutic follow-up for an at-risk patient, request specialized advice, prepare a therapeutic decision, prescribe products, prescribe or perform services or procedures, or monitor the condition of patients.

**Box 1.1.** *Definition of telemedicine in the HPST Law*

Telemedicine is one component of digital health (see Figure 1.1)[2]. The *Conseil national de l'ordre des médecins* (CNOM, French National Medical Association Council), in a report in January 2009 concerning recommendations on telemedicine, states that telemedicine must be distinguished from the notion of telehealth. Telemedicine constitutes a specific subset of telehealth, the main characteristic of which is that it concerns regulated professions. However, telehealth applications are broader than telemedicine applications. This includes, for example, all sites and portals, in whole or in part related to health, that can be found on the internet. These sites, well known to patients and practitioners, offer numerous services: advice, recommendations, articles, forums, newsletters and, for some of them, online medical records.

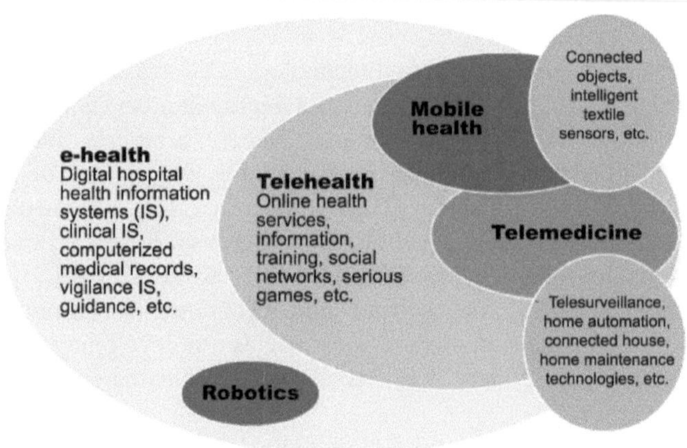


**Figure 1.1.** *The digital health ecosystem*[3]

2 For an analysis of the constraints related to the implementation of e-health systems, in particular the necessary standards and certifications, see [SAL 13a].

3 Source: government report on connected health of the French National Medical Association Council. (Digital health: From e-health to connected health), January 2015, p. 9. In its classic definition, the concept of ecosystem refers to the existence of an "association or community

| Telemedicine stakeholders | Telemedicine companies |
|---|---|
| Software publishers | McKesson, CompuGroup Medical, Corwin, Medisys, Meditrans |
| IT service providers | IT services providers/integrators such as Atos Origin or Sopra Group |
| Manufacturers of medical devices | Philips, Biotronik, Medtronic, Sorin, GE Healthcare, Nx Stage, Physidia |
| Health data hosts | 2CSI, Carestream Health, Cegedim, Cerner, Sigems |
| Telecom operators | Orange, Deutsche Telekom, British Telecom, SFR |
| Telemedicine and telehealth service providers | Intervox, H2AD, Wengo Santé, N2TS, Adechotech, Hailadoc, Teckel Medical, MedGo, Phemium |

**Table 1.1.** *Characteristics of stakeholders involved in the development of e-health in Europe (non-exhaustive list)*

According to SYNTEC NUMÉRIQUE (*Syndicat professionnel de l'industrie du numérique* [French Digital Industry Trade Union]), the telemedicine situation for chronic diseases in France is one of adapted technologies and established regulations, but also an embryonic concretization. After the publication in 2012 of a first white paper (Telemedicine 2020), SYNTEC and SNITEM (*Syndicat National de l'Industrie des Technologies Médicales* [French National Union of the Medical Technologies Industry]) note that chronic disease management is mainly limited to *device management* (reporting alerts on communication equipment) and has not integrated a broader dimension: that of *patient management*. The situation has since changed considerably in the direction of increased psychotherapeutic patient management.

This evolution is primarily due to the actions carried out by various organizations bringing together the knowledge and skills used in telemedicine in France and Europe, including CATEL, the French Telemedicine Society, and *Agir pour la telémédecine*, the European Telemedicine Society (see Table 1.2).

of living beings that interacts with an environment to enable life to be maintained and developed", https://fr.m.wikipedia.org/wiki/%C3%89cosyst%C3%A8me (page consulted on September 12, 2013). The community of living beings here is that of (tele)patients, health professionals and third party caregivers, and the environment represents what is commonly referred to as "the health care system".

| Associations and professional bodies | Missions |
|---|---|
| CATEL – *Club des acteurs de la télésanté* [Telehealth Stakeholders Club] | National e-health reference organization in France. Since 1997, its mission has been to contribute to the development of telehealth, telemedicine and e-health in all its fields of application[4]. |
| *Société Française de Télémédecine* [French Telemedicine Society] | Developing telemedicine research to improve patient management[5]. |
| Société internationale de télémédecine et eSanté [ISFTeH - International Society of Telemedicine and e-health] | Facilitating the dissemination of knowledge and know-how in telemedicine and digital health, as well as access to internationally recognized professionals in these fields of expertise[6]. |
| *Agir pour la télémédecine (Association loi 1901)* [Taking action for telemedicine (Law Association 1901)] | Facilitating telemedicine deployment and collaboration between stakeholders[7]. |
| *Société Européenne de Télémédecine* [European Telemedicine Society] | Stimulating basic, epidemiological, technological and operational research in the field of telemedicine and supporting research programs[8]. |

**Table 1.2.** *Key professional telemedicine organizations in France and Europe*

This phenomenon can also be explained by many factors: the emergence of more efficient and collaborative technologies, the explosion in the number of telemedicine projects, the creation of French Regional Health Agencies (ARSs) that ensure, to a large extent, the coordination of these projects, the deployment of e-health services through the Digital Hospital program piloted by the DGOS, the generalization of the Personalized Medical Record (PMR) led by ASIP health, the *Plan National très Haut Débit* (Very High Broadband National Plan) managed by DATAR at national level and by local authorities and prefectures at regional level and finally the implementation of a National Strategic Plan (NSP) in France for the deployment of telemedicine, piloted by an interministerial committee that has defined five priority projects. These five priorities are as follows: teleimaging, stroke management, prisoners' health, care in medico-social structures or in AHH (at-home hospitalizations) and *management of chronic diseases in the context of remote monitoring (diabetes, chronic end-stage renal failure, heart failure)*. This French National Strategic Plan for the deployment of telemedicine is piloted by an

4 Source: www.catel-group.com/une-grande-histoire.
5 Source: https://twitter.com/sf_telemedecine?lang=fr.
6 Source: https://www.isfteh.org/about.
7 Source: http://www.agir-telemedecine.org.
8 Source: https://www.gralon.net/mairies-france/haute-garonne/association-societe-europeenne-de-telemedecine-et-e-sante-setes-toulouse_W313005083.htm.

enlarged interministerial committee under the chairmanship of the *Direction Générale de l'Offre de Soins* (DGOS, General Management of Healthcare Provision) and including representatives of the following institutions:

– the *Délégation à la stratégie des systèmes d'information* (DSSIS, Information Systems Strategy Delegation);

– the *Agence des systèmes d'information partagés de santé* (ASIP Santé, Agency for Shared Health Information Systems);

– the *Agences régionales de santé* (ARSs, French Regional Health Agencies);

– the *Délégation interministérielle à l'aménagement du territoire* (DATR, Inter-ministerial delegation for land use planning);

– the *Agence nationale d'appuis à la performance hospitalière* (ANAP, National Agency for Hospital Performance Support);

– the *Haute autorité de santé* (HAS, French High Authority for Health);

– the *Association nationale pour la télémédecine* (ANTEL, National Association for Telemedicine);

– the *Assurance maladie* (UNCANSS, Medical Insurance);

– the *Direction générale de la compétitivité de l'industrie et des services* (DGCIS, General Management of Industry Competitiveness and Services);

– the *Direction générale de la cohésion sociale* (DGCS, General Management of Social Cohesion).

This plan is structured around three components: a 2011/2015 strategic deployment document, an action plan and a toolbox. The management of chronic diseases as part of telemonitoring is the central focus of the research and reflection carried out in this book. Telemonitoring of chronic diseases is one of the five telemedicine procedures that can be carried out in France defined in the 2009 HPST law (see Table 1.3).

Telemedicine makes it possible to establish a diagnosis, to ensure, for an at-risk patient, a follow-up within the framework of prevention or a posttherapeutic follow-up, to request specialized advice, to prepare a therapeutic decision, to prescribe products, to prescribe or carry out services or acts and to carry out a monitoring of the state of the patients. Its challenges are clearly defined as follows:

– to reduce as much as possible certain inequalities in access to healthcare, particularly for isolated patients, whether they are living in isolated areas (rural, mountainous, island, etc.)[9] or people in detention;

9 See [SAL 13b].

– to provide an organizational and technical response to the numerous epidemiological (increase in the number of patients suffering from chronic diseases and/or polypathologies linked to the aging of the population), demographic (unequal distribution of professionals on the national territory) and/or economic (budgetary constraint) challenges facing the health system today.

| | |
|---|---|
| *Teleconsultation* | A physician gives a remote consultation to a patient, who may be assisted by a health professional. The patient and/or the professional at his/her side provides the information, and the remote doctor makes the diagnosis. |
| *Tele-expertise* | A doctor remotely seeks the opinion of one or more colleagues on the basis of medical information related to the management of a patient. |
| *Remote medical monitoring* | A physician remotely monitors and interprets a patient's medical parameters. Data recording and transmission can be automated or performed by the patient him/herself or by a healthcare professional. |
| *Remote medical assistance* | A doctor remotely assists another health professional during the performance of an act. |
| *Medical regulation* | The doctors at the emergency centers make an initial diagnosis by telephone in order to determine and trigger the response best suited to the nature of the call. |

**Table 1.3.** *The five telemedicine services available in France*

In order to facilitate access to patient care, French Ministry of Health encourages the development of telemedicine in the city and the medico-social sector. This innovative form of healthcare organization is still too limited to the hospital field. It is therefore necessary to remove the obstacles to its development. It is for this reason that the 2014 Social Security Financing Act provides a framework for experimentation with the financing of telemedicine acts in several pilot regions: in towns (homes and health centers, private practices) and in medico-social structures (retirement homes, EHPAD). Thus, nine regions have been selected and are experimenting with prefiguring and incentive tariffs. These regions are Alsace, Basse-Normandie, Bourgogne, Centre, Haute-Normandie, Languedoc-Roussillon, Martinique, Pays-de-la-Loire and Picardie.

The first patients treated by telemedicine will benefit from remote consultations by specialist doctors, whose distribution is sometimes uneven throughout the region. Telemedicine offers geriatric and psychiatric care by doctors working in hospital centers, sometimes far away, to certain patients with complex pathologies, such as elderly persons residing in residential institutions for dependent elderly people (EHPAD, see Box 1.2). These experiments offer the possibility of improving the follow-up of patients at home, thanks to remote monitoring, in particular for those suffering from heart failure, renal failure and/or diabetes.

**EHPAD**

Residential homes for dependent elderly people (*établissements d'hébergement pour personnes âgées dépendantes*, EHPADs) are nursing homes. They are aimed at people over 60 who need help and care on a daily basis. The EHPAD's mission is to accompany frail and vulnerable people and to preserve their autonomy through comprehensive care including accommodation, catering, entertainment and care. The EHPADs sign an agreement with the departmental council and the ARS (French Regional Health Agency), which provide them with funding in return for quality care objectives.

**Box 1.2.** *Definition and missions of EHPAD*[10]

Article 118 of the Hospital Reform, Patients, Health and Territories Law of July 21, 2009 also created the French Regional Health Agencies (ARSs). Their mission is to ensure unified health management in the regions, better respond to needs and increase system efficiency (see Box 1.3). The strategy of an ARS is defined in its Regional Health Project (PRS). ARSs are the pillar of health system reform.

**French Regional Health Agencies (ARSs)**

The regional health agencies are public institutions, morally and financially autonomous, placed under the supervision of the ministries in charge of social affairs and health. They are responsible for the regional management of the health system. They act in the field of healthcare as a whole:

– prevention and health promotion;

– health and safety monitoring;

– organization of outpatient care services (local medicine) and health institutions (hospitals, EHPAD);

– organization of medical and social care management.

**Box 1.3.** *Definition and missions of ARSs*[11]

10 Source: National information portal for the autonomy of the elderly and the support of their relatives, http://www.pour-les-personnes-agees.gouv.fr/choisir-un-hebergement/vivre-dans-un-etablissement-medicalise/les-ehpad. Page consulted on December 6, 2017.

11 Source: website (Regional Health Agency). https://www.ars.sante.fr/quest-ce-quune-agence-regionale-de-sante. Page consulted on December 6, 2015.

Twenty-six ARSs have been created in France. Those that currently have a Regional Telemedicine Programme (RTP) are as follows: Alsace, Auvergne, Basse-Normandie, Bourgogne, Centre-Loire, Franche-Comté, Guyane, Haute-Normandie, Île-de-France, Languedoc-Roussillon, Martinique, PACA, Picardie, Poitou-Charente and Rhône[12]. In addition, nine ARSs were selected to conduct *telemedicine pilot projects*. This appeal was confirmed under the Social Security Financing Act for 2014, via the launch of trials in nine regions of new telemedicine acts to combat medical access deserts in the outpatient and medico-social sectors.[13]

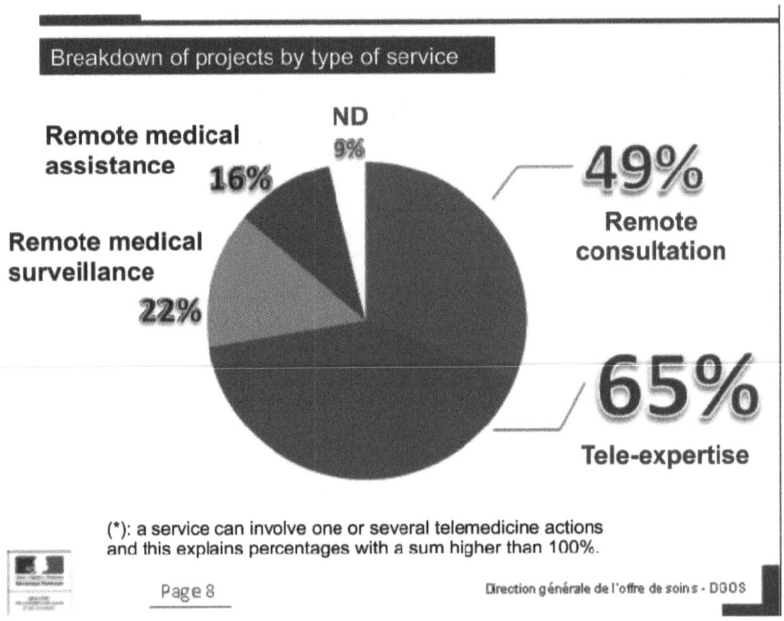


**Figure 1.2.** *Types of telemedicine performed in French ARSs [DGO 13, p. 8]. For a color version of this figure, see www.iste.co.uk/bardy/telemedicine.zip*

---

12 In line with national priorities, however, 25 out of 26 regions have at least one telemedicine device for the management of a chronic disease – mainly kidney failure, heart failure and chronic wounds – and 30 telemedicine devices for the management of stroke are identified in 23 regions. Moreover, although the hospital sector still predominates (53% of the devices listed are exclusively hospital based), 30% of the devices now involve the outpatient sector (95 devices, 51 of which are operational). Source: www.ars.sante.fr. (page consulted on September 20, 2014).

13 Nine pilot ARSs to test the remuneration of telemedicine acts within the framework of article 36 of the Social Security Finance Act (LFSS) 2014 are as follows: Alsace, Basse-Normandie, Bourgogne, Centre-Val de Loire, Haute-Normandie, Languedoc-Roussillon, Martinique, Pays-de-la-Loire and Picardie. Source: www.ars.sante.fr. Page consulted on September 20, 2014.

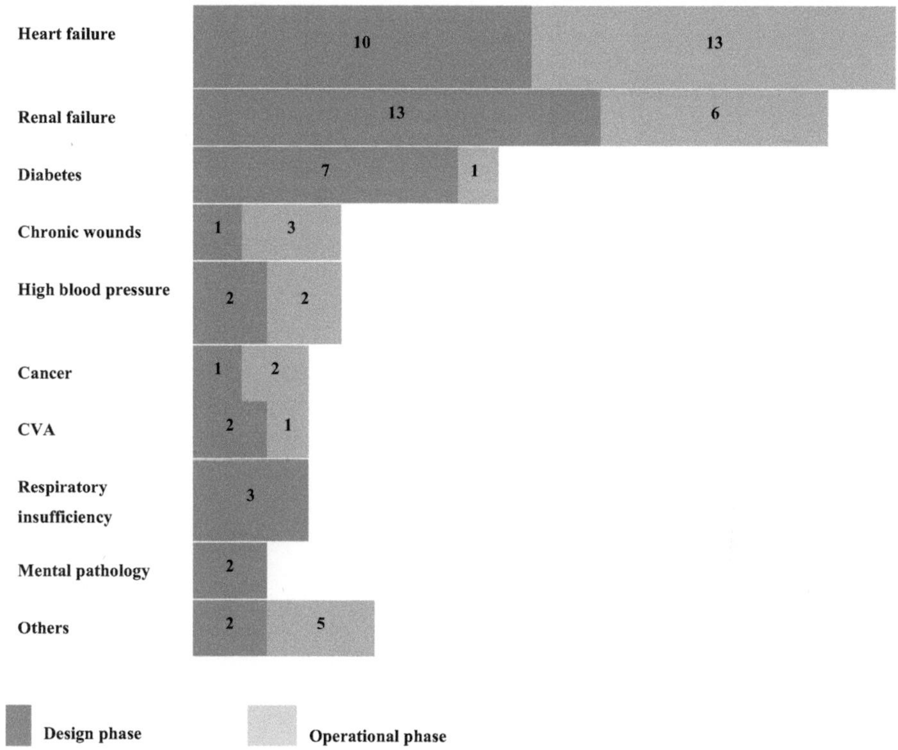


**Figure 1.3.** *Medical telemonitoring projects identified by DGOS [DGO 13, p. 9]*

According to estimates by DGOS, 78% of telemedicine projects involve teleconsultation and tele-expertise. In 2013, the DGOS also identified 76 remote medical surveillance projects. A 20% increase in the total number of telemonitoring projects in 2015 and beyond can be estimated, particularly in the area of chronic diseases (see Box 1.4).

**Key figures**

331 telemedicine projects in 2012, including 30 in Île-de-France

Nine pilot regions

The distribution of telemedicine projects according to four chronic diseases mainly concerned is as follows:

– heart failure, 26%;

– renal failure, 32%;

– chronic wounds, 20%;

– diabetes, 16%.

**Box 1.4.** *Breakdown of telemedicine projects in France in 2012. Source: [DGO 15, p. 20]*

Three quarters of the telemedicine projects identified fall within the five national priorities. The three top national priorities are chronic disease management (26%), imaging as part of ongoing care (24%) and stroke management (10%). The other two priorities, namely the health of prisoners and care in a medico-social structure or home hospitalization, represent 5% and 11% of activities, respectively. Of the 66 chronic disease management activities, renal failure (26%), cardiology (21%), cancer (16%) and diabetes (8%) are the most affected. Thus, a quarter of the activities are outside of these priority projects. This most often involves imaging outside of permanent care (27%).

Projects currently under design are oriented more toward a regional deployment strategy than toward projects that are already operational. Thus, 51% of operational activities have a fairly local scope, limited to a few health professionals or organizations, without strong institutional and organizational support. However, 39% of them have an extension strategy. Projects under design are regional in scope from the outset: only 16% of them have a strictly local scope. Telemedicine activities are essentially based on the resources of health establishments: 57% of projects are exclusively hospital based and 85% concern at least one health establishment. Of these, the public sector is predominant (83%).

Among telemedicine actions, those of tele-expertise are the most often developed. They concern 48% of the projects monitored by teleconsultation acts (25%), telemonitoring acts (16%) and teleassistance acts (10%). Telemedicine therefore occupies, as we can see, a preponderant place within the so-called *connected health* ecosystem that includes, at least in its current sense in France, telemonitoring, home automation and even so-called "mobile health" (mhealth). The choice was made here to focus on telemedicine research, and more particularly on telemonitoring coupled with autonomous home care, or telecare, because this is actively utilized in the treatment of chronic diseases. Treatment of chronic diseases represents a major societal challenge given the number of people concerned in France, and it is part of a sector with strong growth potential where economic models and beyond, i.e. the socio-philosophical analyses that accompany them, remain to be built.

## 1.2. Telemedicine: medicine without patients?

Philippe Thieblot, President of the *Conseil régional de l'ordre des médecins* (Regional Medical Association Council), pays particular attention to the application of digital health, ensures compliance with medical ethics in its use and launches a public warning in 2013:

> *These practices could make a lot of money without patients benefiting from them, through telemedicine clinics. With the uses, we thus remain very vigilant to avoid any deviation*[14].

The economic logic of contemporary telemedicine, between reducing costs and supporting the development of telehealth participants, could give rise to fears of the advent of impersonal and commercial medicine favoring a strategy of profitability over a humanistic approach to care. This risk is perfectly conceivable within the framework of medical care, which would remain at the stage of care management via interposed devices (*device management*). However, as has been pointed out, 2014–2015 represents a pivotal year in the development of telemedicine in France due to the diversity and scope of the initiatives carried out in telemedicine (notably by the health authorities). The year 2014–2015 can be seen retrospectively, for this reason, as "year 1" of medical care refocused on *patient management*. The risk of a commercial drift in telemedicine and/or impersonal and technical patient management, although still possible, therefore seems less likely, especially since patients, federated within associations after several years of experimentation, now make their voices heard clearly and assert their rights. Computer-assisted medical care has therefore come to a standstill and is now giving way to medicine in which the use of devices gives patients the status of "participants in their own health" and their experience has new value, as digital health consultant Yves Jarlaud points out in a recent article:

> *Taking into account the patient experience in long-term conditions, i.e. the different dimensions of the patient's life with their illness and their treatment, is an essential determinant of the quality of the care pathway and is now at the heart of the reflections and concerns of all participants in the healthcare chain. This is also true for pharmaceutical companies, which are aware of the growing influence of patients in prescriptions, integrating patient experience at a very early stage in the design of innovative therapeutic solutions likely to improve the compliance and effectiveness in real life of the treatments they develop*[15].

---

14 2013 activity report of the Auvergne ARS, p. 48.

15 [JAR 15]. www.hospitalia.fr/Rendez-vous-aux-3emes-Assises-de-l-Innovation-Therapeutique-Le-parcours-de-soins-a-l-heure-de-l-experience-patient_a712.html. Page consulted on November 10, 2015.

As Yves Jarlaud points out here, the effectiveness of telemedicine treatments depends, and will depend more and more, on the management of the patient's experience, as well as on the patient's involvement as a co-participant in the care required to master new tools to manage their health. In a recent book, Nelly Oudshoorn, a health technology specialist, explains the reasons that lead the patient to assume this responsibility: "The actions of patients involved in monitoring and producing diagnosis can thus not be considered merely in terms of performing tasks, particularly if this implies the adoption of a purely functional perspective. Telecare technologies do much more than act as functional tools to monitor data; they transform the relationship between patients and their bodies. This is because patients are expected to become more active and responsible as participants in the diagnosis and monitoring of their disorders"[16].

The telemedicine patient, renamed "*interactive patient*", is therefore evolving in a singular context at the crossroads of technological progress and the human. For Yves Jarlaud:

> *This historical conjunction between the fruits of scientific research and technological progress will obviously have direct consequences on medical practice, hitherto mainly based on the action given by a health professional to a passive patient. In this new context, it is increasingly evident that the effectiveness of treatments will also depend on the quality of interactions within the 'healthcare ecosystem', and, in particular, on the patient's adherence and participation that the social media transform into an 'interactive patient'. A patient who will also see the relationship to their body and disease modified by the generalization of self-measurement, therapeutic information, and/or adherence programs. We are thus moving from passive consumption of health products prescribed by the doctor to the 'experience' provided by an increasingly personalised therapeutic solution deployed as part of an 'intelligent care pathway'. One more step towards an accompaniment-based medicine to which virtual patient communities are already making a spectacular contribution*[17].

Telemedicine is therefore not medical care without patients but a significant evolution of care practices, or even a "revolution in the 'revolver' sense, the Latin etymology of the term, that is to say, a return to medicine based on the relationship and therefore on the human!" [JAR 14]. Bruno Salgues, director of studies at Institut

16 [OUD 11, p. 10]. See also [LEH 04].

17 [JAR 14]. http://technologiesnumeriquessante.aromates.fr. Page consulted on March 20, 2014.

Mines Telecom, and author of a reference work on the industrialization of health, also stresses that “the current situation is dominated by the patients’ claims to a more dominant role within the health ecosystem [...] It is only recently that the notion of ‘quality of life’ has taken root in the medical field” [SAL 16, p. 25]. With the advent of “industrialized” medical care, which cannot ignore the qualitative dimension of care, a new figure of the patient is asserting itself: that of the participant of their own health or “health actor” (“*santacteur*” in French, to use here the neologism forged by Bruno Salgues) [SAL 16].

We can certainly agree on the fact that telemedicine seems to have redefined of patients support in France since 2014. Experience shows, however, that the human management of patients in general and chronic patients in particular, is still far from being achieved. Health authorities are probably aware of this problem if we are to believe the emphasis placed in telemedicine today on long-term conditions (LTC). It should be noted that in France, unlike some English-speaking countries (including the United Kingdom and the United States), LTC is designated as a condition that gives rise to full reimbursement of care, after establishment of a therapeutic protocol. In France, there are 30 LTCs on the list drawn up by the French health insurance scheme[18]. The reimbursement of LTCs is not fully covered by health systems in some other European countries or in the United States.

## 1.3. Telemonitoring of chronic diseases: a national priority

QUESTION.– What is remote medical surveillance?

Medical telemonitoring is a continuous or non-continuous monitoring process that allows a healthcare professional to remotely interpret the data necessary for a patient’s medical follow-up and, if necessary, make decisions regarding the patient’s state of health (for example in the case of remote monitoring of a cardiac patient with a defibrillator). Home telemonitoring of chronic patients is a predictive mode of disease management specific to telemedicine. This type of telemonitoring is also a national health priority.

The objective of remote monitoring coupled with autonomous care at home, carried out by chronic patients at home, is to enable people to live at home for as long and independently as possible, in an environment of comfort and safety. It is about detecting and preventing the occurrence of critical situations at home or deterioration in a person’s state of health. These systems thus represent a temporary or lasting alternative to hospitalization or the use of long-term residential facilities – retirement homes or specialized centers. The patient is therefore no longer forced to

18 https://fr.wikipedia.org/wiki/Maladie_de_longue_durée.

give up their home and life in society. They retain a large degree of autonomy in their social and private environment, while benefiting from preventive health services. These systems concern the elderly, but more generally people who are at risk of motor (e.g. falls) or cognitive disorders (depression, senile dementia, etc.), or who require special care or attention (diabetics, renal failure).

## 1.4. Specific features of telecare

One of the specific features of remote monitoring is the need to process data rapidly over time in order to meet the objective of detecting situations of concern at home as quickly as possible. The difficulties of these analyses are linked to the heterogeneity of the data collected, the influencing factors acting on the observed parameters, and the mutual dependencies of these parameters.

We now will propose here to clarify the terms used in this research. A nuance exists between the notions of *telemonitoring* and *tele-follow-up*: the first refers to the practitioner's view of the patient while the second, conversely, refers to the act as it can be perceived by the patient. In addition, telemonitoring appears to correspond, *in practice*, to overall patient monitoring where tele-follow-up appears to focus more specifically on the patient's physiological data. However, the terms *telemonitoring* and *tele-follow-up* may be used interchangeably in the following chapters, in a generic sense to mean *the medical control exercised over the practice of home care*, regardless of the patient or the physician's point of view. On the other hand, remote monitoring refers to the use of communication devices and is therefore distinct from autonomous care at home, which relies on the use of specific care devices (for example a computerized insulin pump or a dialysis cycler). The term "telecare" therefore refers here jointly to these two categories of medical devices and practices (care and telemonitoring). In other words, *telecare refers to a (co-managed) medical practice that is inseparable from a communication practice.*

This definition of French telecare differs from the meaning generally attributed to its English equivalent. In English-speaking medical culture, the term *telecare* refers to remote care provided to dependent elderly people and preventive home automation solutions implemented at home (sensors and cameras). The concept of *telecare* mainly refers to the empowerment capacity of dependent people (diabetic patients are not generally considered as the target audience of *telecare*). However, there is no consensual definition of *telecare* today: its polysemic nature is therefore accepted.

As defined above, telecare allows patients with chronic illnesses or disabilities to perform their own home care duties – such as blood pressure monitoring – and share this information electronically with their physicians and other health care providers. The medical team monitors patient outcomes on a regular basis and adjusts treatment as needed. In the long term, telehomecare serves as a guide for patients to help them live with their chronic illness.

The telecare solution offers patients access to an application via a tablet or personal computer that can be used at home or outside the home (in the case of the diabetic patient telemonitoring). The intuitive application guides patients through each step and is based on predefined questions that they must answer (variable according to the pathology). In some cases, this process may occur several times a day (for patients with type 1 diabetes or patients on home dialysis). Telecare also allows the use of medical devices (connected blood pressure monitor or scale) to collect health data and transmit it to the health care provider and medical team. The telehomecare solution provides healthcare professionals with access to a centralized view of all patients participating in the telehomecare program by allowing them to tailor their work methods, protocols and interventions to create personalized care plans based on patients' health condition or status. Telecare also generates alerts and reminders based on forms created by clinicians or data obtained from patients (for example high blood sugar). The use of a telehomecare solution thus contributes to increased patient satisfaction and overall quality of care through closer interaction with health care professionals, reduced anxiety and fewer emergency-room visits or hospital stays. Patients also appreciate receiving care at home rather than in a hospital.

On the basis of this observation, the triple phenomenological, epistemological and ethical analysis of telecare, proposed in this book, aims to shed light on the remote monitoring of chronic diseases that cannot be approached from the sole objective and/or functional angle. Remote monitoring of chronic diseases can also be seen from the point of view of culture (beliefs, values) and the patient's psychoaffective dimension (physical and psychological experience of the disease) as well as from singular biographical elements (memories, family and personal history, anticipation of the future), because of which the patient can forge a *human* therapeutic alliance with the care givers, find their way in the world and interact with others. In other words, a new knowledge of chronic disease and its multiple temporality is being developed today within the framework of telecare whose relational dimension, calling for particular attention, constitutes – in addition to self-managed care – a major ethical challenge for contemporary telemedicine: far from the frightening perspective of the topic shaped by a technologizing medicine.

# 2

# The Human Ethical Challenge

*No general code of ethics can*
*tell you what you ought to do; there are no signs in this world.*
Jean-Paul Sartre[1]

We prefer the term "ethical" rather than "moral" here, based on the distinction proposed by psychologist Clotilde Leguil:

> *The term ethics differs from that of morality in the sense that morality refers more to a body made up of norms, whereas ethics implies a questioning of the norm itself. Ethics questions the foundations of these norms and at the same time is confronted with the absence of immutable moral criteria.* [LEG 08, p. 3]

As the ethicist Jérôme Béranger points out, "Ethics […] involves calling into question the values which underpin action […] Ethics is an individual tendency to act honestly in a given situation, so as to make the right decision" [BÉR 16, p. xxi]. This ethicist will also be followed in their analysis of "value" and ethics applied to digital:

> *This word [value] has a general and dynamic connotation. It has, primarily, a philosophical evocation before having an ethical consequence. One of the foundations of ethics is this imperative to appeal to actor rationality. This idea is achieved through an understanding in coordination, exchange and sharing between the protagonists. Each person contributes to seeking cross-comprehension of the situation analyzed. This, therefore, presupposes a certain*

1 [SAR 72, p. 47].

*consensus and solidarity between the interlocutors who share the same aims. If ethics is always, by its very nature, complex to define, putting it into perspective with digital technology constitutes another challenge. Ethics demands a vision, a design and an ambition, which takes shape in a given direction.* [BÉR 16, p. xxv]

At the dawn of the 21st Century, ethics is built in the confrontation of individuals, free and equal in law, with the new possibilities offered by science and technology that sometimes seem to threaten human respect, dignity and freedom as witnessed by the creation of ethics committees made necessary by the rapid advances in science, in general, and in medicine and the life sciences, in particular. In addition, as the philosopher Bernard Reber points out in a recent book, the 1990s added the need to *include* civil society and *involve* various stakeholders in ethical decisions:

*The hegemony of assessments performed by experts, ethics committees or analysis from social sciences researchers has become an object of criticism. What has been challenged is not so much the moral validity of efforts that these mechanisms mobilize, but the fact that they emanate from an overhanging perspective, impervious to how ordinary citizens assess technologies or research. For this reason, these experts' analyses are now seen as insufficient resources to appraise the social acceptability and ethical desirability of science and technology. Since ethics committees only inadequately represent and reflect values and value systems of the members of society, they could not be the only source of fundamental normative decisions.* [PEL 16, p. xiv]

At the same time, as the philosopher Bernadette Bensaude-Vincent points out, the involvement of citizens in societal choices implies a new configuration of the relationship between science and opinion:

*As long as you hope to engage the public from a managerial approach, you don't really requalify opinion, you subject it to economic rationality. To the extent that opinion is knowledge derived from practice, and adapted to the local, it would be necessary to value situated knowledge, without claim to universality, knowledge developed in research-actions circumstantiated*[2].

Symptomatically, contemporary ethics is interested in the question of limits: Those that should be jointly imposed on science and technology in their relationship

2 [BEN 13a], postface [digital format]. Also see [BEN 13c].

to the human, and on the subject, whose power to act is technologically increased. As Clotilde Leguil still writes:

> *The human body is not a commodity, suffering is not a dysfunction, the child is not a manufactured object, the psyche is not a computer. Otherwise there can be no possibility of ethics.* [LEG 08, p. 6]

These premises raise three essential questions for reflection on contemporary telecare: From what ethics does the relationship between new health technologies and chronic patients today emerge? What is ethical telecare? What does the concept of *person* mean in a telecare context?

Answering these multiple questions involves reexamining, on the one hand, *the underlying ethical background of telecare* and circumscribing *the field of influence of medical ethics* from which it can be understood, and then the clarification, on the other hand, of *the specificities of the person* within the framework of a technicized care relationship. This three-pronged approach constitutes an introduction to put into perspective the ethical analysis of telecare in its relational, temporal and psychoaffective dimensions, which will be conducted here, through the prism of care ethics[3].

Let us note in passing that *care*, that is to say care as the attention to the other, has been opposed to another category of care since the 1970s, the *cure*, that is to say care in its technicality, as witnessed by the analysis of the psychoanalyst Donald Winnicot. For Winnicot, "Cure, in the sense of treatment, in the eradication of the disease and its cause, tends (...) to take precedence over the sense of care (care, interest, attention)". However, the categories of care and cure, in the context of telemonitoring, are not in opposition: telemonitoring technologies for chronic diseases are not indifferent to humans, just as human care is not resistant to telemonitoring technologies. Nor is telecare a practice aimed at restoring the precedence of the cure on the care, or the reverse (as Winnicot suggested). For Winnicot, as the philosopher Xavier Guchet points out, "once the reality of the indisputable contributions of science and technology to medicine has been reaffirmed, the philosopher can deal with the essential, i.e. the problem of how the doctor takes care of their patients by considering them as singular suffering subjectivities, and not only as object bodies accessible to the knowledge and powers of medical techniques" [GUC 16, p. 18]. Winnicot suggests, through this approach, to restore the procedure of the care, threatened according to him, on the cure.

However, one of the aims of this work is to show how the categories of care and cure, in the context of telecare, enter into a relationship of codependence, and how

3 See Chapter 8.

an applied *care ethics* makes it possible precisely to identify the specificities and the key issues.

The emphasis here is, therefore, on the ethical meaning of *care* and *cure* and not on their meaning in terms of the health system. In the case of LTC, the *care* and *cure* are treated differently according to the system: financial aid from the state co-managed by the patient in a universal Beveridgian system as in the United Kingdom (personal health budgets), or joint care and cure by health insurance in a Bismarckian system (France, Germany).

## 2.1. Ethical background of telecare

Telecare as a technological *relationship*, a communication *device* and *medical practice* co-managed by the patient and the practitioner/medical team first of all involves ethics because of its technological dimension. The major ethical texts (UN, UNESCO) address the issue of patients' vulnerability linked to technologies: Whether it concerns the handling of communication devices or the complexity of remotely digitized protocols and the risks of privacy and intimacy invasion that they presuppose. Telemonitoring programs – these being only one dimension of telecare – are also perceived as expert systems that raise the question of the place of the physician and the patient within the protocols. The fears expressed about telemedicine also concern the overuse of connected devices around the patient with chronic disease and the risks of objectification of the person via the acquisition of skills transforming them into a *digital patient*.

In addition to these technical difficulties, there is the issue of medical information. There is discrimination through information, whether it is access to the personalized medical file or follow-up medical data collected through communication devices.[4] However, medical information is not undifferentiated information: it concerns the intimacy and integrity of the person and refers to representations. Ethics also involves reflection on how patients and doctors appropriate the knowledge necessary for good quality therapeutic follow-up.[5] In the context of telemonitoring – and this point is often raised in studies focusing on medical practice – the knowledge and competence of the physician go hand in hand, in the eyes of patients, with the desire to treat the person in a way that is not protocolized or proceduralized. The major ethical issue raised by going beyond a "technical" approach to care is that of the values of *living together* and *solidarity*.

4 The notion of transparency, often mentioned in ethical debates, also remains relative because it refers to the patient's ability to decrypt medical information.

5 This question refers in particular, as the testimonies gathered from patients for the purposes of this book show, the need to strengthen therapeutic education and to define precisely the place it should occupy within the framework of telecare.

However, the care relationship must not be overwhelmed by medical information and living together must not be eclipsed by, or reduced to, "managing care together" ("together-management")[6].

Aristotle's invitation to prudence in *Nicomachean Ethics*, as a provision for deliberation on what should be done (according to what is deemed good or bad), finds a favorable echo in the context of telecare in what is expected of the physician: to treat the disease with respect for the patient and their needs, in particular the need for empathic listening. The care relationship, and telecare, implies establishing a partnership that is both prudent *and* fair between the participants concerned (in the sense of an equitable sharing of available resources): not to abuse expert knowledge that is indifferent or insensitive to the person's experience.

The ethical background of telecare, outlined here, is also inspired by the four ethical principles defined by Beauchamp and Childress (2001)[7], summarized by Jérôme Béranger in a recent book.

– Autonomy

This designates the fact that an individual allows themself to have their own rules of behavior, as the Greek terms "*autos*" and "*nomos*" respectively mean "self" and "law or rule". The purpose of this principle is to involve the patient in the decision-making process.

– Beneficence

Beneficence contributes to the well-being of others. It must fulfill two highly precise rules. The action undertaken must be beneficial and useful, that is to say have a positive cost-benefit relationship.

– Non-maleficence

This aims to avoid harm to the person to whom you owe the responsibility (the patient) and save them from the harm or suffering which would make no sense to them. Its aim therefore, is both to do good and abstain from doing harm. This principle appears in the Hippocratic maxim *primum non nocere*[8], the consequence of which is to do good to patients and to stop them from being harmed and subject to injustice.

6 This term is borrowed from socio-anthropologist Jeannette Pols. It designates the possibility for doctors and patients to "technically" co-manage care and share patient data without their psychoaffective dimension being really taken into account. See [POL 12, p. 75].
7 See [BEA 08].
8 "First do no harm".

– Justice

Justice aims to share available resources between all patients [time, money and energy resources]. This principle is strictly linked to the notions of equality and equity, which play a part in the process of making fair decisions directly. Ideally, all actions may tend towards a perfect equality, but according to the circumstances and the nature of the individuals involved, equity is often essential both to establish priorities and a particular hierarchy in the actions to complete. This principle includes a scope concerning all patients which may be designated as "macro-ethic" while the three previous principles have a far more individual and relational dimension and may be considered as "micro-ethic".

**Box 2.1.** *Vocation of the four ethical principles [BÉR 16, p. xxiii]*

The principles of autonomy, beneficence, non-maleficence, and justice, pillars of any ethics applied to telecare, must allow practitioners to accompany the patient's ongoing negotiation with their chronic illness. As the testimonies collected for this book indicate, it is indeed essential to take into account the fact that the patient's ""truth" about their illness is at once *subjective*, *complex*, and *evolving*. In this context, the role of the physician is to give meaning, coherence, intelligibility and direction to the patient and to create the conditions for trust and living together. Allowing the person their privacy is an indispensable part of telecare. Faced with the supposed dangers of disempowerment (telemedical technologies would ensure the work of *taking care* in the place of the patient or doctor) and hypersurveillance (these same technologies would threaten free will and the freedom to be oneself), communicating and interacting within the framework of telecare, respecting the principles of autonomy, beneficence, non-maleficence and justice, thus implies a commitment not only on the part of the caregiver but also on the part of the patient being cared for.

DEFINITION 2.1.– What is *compliance*? Behind the ethical problem of self-actualization, the necessity of the patient's adherence (and their limits) can be seen[9], their consent, what is sometimes referred to as *compliance*.

The term *compliance* refers here in concrete terms to the acceptance of telecare standards: on the one hand, technical and medical standards imposed by care devices (measurements of blood sugar levels and heart rates at regular intervals according to

9 On this point, see the work of Nelly Oudshoorn and Ivo Matthuis who shed socioanthropological light on the recalcitrance of patients suffering from chronic respiratory insufficiency as part of the recent clinical evaluation of a remote monitoring protocol based on standardized monitoring methods [OUD 15].

a predefined protocol), relational and/or communicational standards (exchanges of information between participants) and, on the other hand, a certain form of discipline in the self-managed practice of home care, or even conformity, which is part of a long-term process. These various aspects of compliance are addressed, from different angles, directly or transversally throughout this book[10]. An ethical practice of telecare also imposes on health professionals and their teams responsibilities toward the patient, often new, including the need to respond to the empathic demand of the (tele)patient, but also to create the conditions for a human approach to telecare: that is, a collegial decision-making process, a dialogue that allows the person and their family to be integrated from the outset, rather than the transmission of information as and when it happens.

The questions raised here by telecare involve ethics, but ethics adapted to specifically technomedical practices. To date, however, there is no ethic of telecare as such, in the sense given to this term in this book. Nevertheless, it seems that medical ethics can provide valuable analytical tools in this field.

## 2.2. Circumscribing the field of influence of medical ethics

In a context of ethical subdetermination, the technologicalized care relationship that engages patients with chronic diseases and physicians can indeed rely on medical ethics, because telecare, like care, is confronted with the same expectations of humanization of practices, uses and the caregiver–patient relationship.

### 2.2.1. *Communication between doctor and patient*

The medical relationship, as the philosopher Emmanuel Hirsch stresses, cannot be reduced to the quality of the transmission of information:

> *If we reduce our thinking today to the medical relationship, in other words if we limit it to the quality of the transmission of information, we are out of the game. The spirit of the law on bioethics of March 4, 2002, is to give the person the opportunity to appropriate knowledge within the framework of a partnership that allows them to make the fairest decision and to be recognized in their faculty to decide, to accept, and to refuse. We are in the very foundations of contemporary biomedical ethics with the Nuremberg Code, article 1 which is on*

10 See, on this point, Chapters 4 and 5.

*consent, and the real question, in consent, is that we must inform in an objective manner, without bias, without abuse.* [HIR 12]

In other words, it requires a real act of communication. To succeed in this act, the doctor must be able to speak to the *patient-as-subject* about the *body-as-organism* that is part of clinical know-how. This question, this paradox itself, is complicated in the context of telecare since communication between the participants concerned, digital by nature, favors procedure (and therefore technical know-how) over speech (as the ability to express nuance). Even more than care, telecare thus requires the learning of a common language, comprehensible by all participants, through which suffering and malaise can be heard beyond the technical narrative of the disease.

The sick body is not an *object-body* but a *lived body*. As Maurice Merleau-Ponty states in *Phenomenology of Perception* (*La phénoménologie de la perception*): "The corporeal event always has a psychic meaning" [MER 15, p. 187]. The doctor must respect the psychological meaning of the illness for the person, but also know how to give it meaning. The art of communication, and the empathy of the doctor[11] , is thus at the crossroads of the body and the psychic: the doctor allows the patient to regain power over what happens to their body and to live with it. However, this practice of communication is not scientific: medical practice is always based on the health professional's judgements and therefore on their subjectivity. In other words, the physician's knowledge is partly subjective since it is based on an interpretation that cannot be translated into measurable data. In the telecare context, however, the physician's subjectivity poses a double problem.

On the one hand, they often respond to a request for empathy from the patient. However, the physician's propensity to respond to this expectation quickly comes up against the limits of their prerogatives: it is a matter of treating patients and not entering into sympathy with them. As for the empathic response, it cannot always live up to patients' expectations; doctors themselves recognize their limits in this field because of the multiple constraints exercised by their profession (limited time for consultation, priority of the medical dimension of care). Empathic listening is increasingly delegated, in the context of telecare, to nurses and psychologists. The telecare protocols that have proved their worth in France to date have all integrated strong psychological support for the patient, outside the patient–physician relationship.

Since the law of March 4, 2002, the principle of transparency and the right to information have also risen up against the empathic deployment of the medical team toward chronic patients. Now the patient has the right to be informed. The medical

11 See on this point the following reference work: [HAL 01].

decision implies free and informed patient consent, but also the right and duty of the physician to enlighten the patient. As the philosopher Jacqueline Russ points out, the 2002 law profoundly modifies the exchange between doctor and patient and medical practice by, however, removing the physician from their position of authority:

> *Physicians must comply with standards imposed on them. The medical decision was somehow "desubjectified" to make it a decision based on impersonal scientific rationality. The medical decision is therefore no longer purely an individual decision, but the result of crossing several data resulting from a protocol. This approach comes from the industrial world and refers to a system of standardization of practices. Every physician must make the same decision in the same clinical situation. It is the quality approach that aims to "optimize" care, that is, to make medicine more effective, more profitable, by standardizing practices. Thus the medical decision, increasingly anonymous, must then be evaluated objectively. It is not a doctor who treats a patient, but a protocol that is put in place to restore health. What can then be blamed on a doctor is not their lack of tact, their diagnostic error, but their failure to follow the protocol*[12].

Faced with the physician's difficulty in exercising their subjectivity, which is nevertheless necessary for the well-being of the person in the context of telecare, it seems useful to summon here an ethics that makes it possible to frame empathetic telecare practices.

This ethical issue does not only concern telemedicine participants in France. It is also shared in many countries in Europe and throughout the world. A recent in-depth study of the telemedicine literature for the period 2012–2017 thus reflects a common demand: that of providing practitioners with standards and guidelines aimed at preserving the relational dimension of care and, more generally, at guaranteeing quality telemedicine services to patients[13]. This study corroborates the 2014 European guidelines for telemedical care, which state that the services offered "must take account in a timely manner of the advice, opinions, and choices of the caregivers and the cared for" [TEL 14, p. 28]. As a follow-up to initiatives undertaken in Europe to combat the risk of dehumanization of the caregiver–patient relationship, the American Medical Association has issued ethical guidelines in

12 See [RUS 94].

13 See [LAN 17]. Based on several major databases (PubMed, Scopus, Science Direct), the authors have identified 503 relevant scientific articles, for the period 2012–2017, relating to the ethical issues of contemporary telemedicine.

2016 also aimed at ensuring good practices and "personalized" use of new telemedical technologies in the United States[14].

### 2.2.2. *The contribution of care ethics*

Care ethics is thus distinguished from medical ethics by the fact that it "emerges as the discovery of a new morality whose voice must be recognized in a world that does not have the adequate language to express and have recognized all that is part of the work of 'taking care of the patient' and the feeling of responsibility towards the well-being of others" [BRU 11a, p. 7]. Care ethics was conceived, from its origin, as a relational ethics structured by attention to others, to the other than oneself. It thus joins the new approaches to the telecare of patients with chronic diseases, which today tend to value a *respectful* work "of taking care of the patient" *shared* by the participants of the technicized relationship, in its double therapeutic and human dimension.

Care ethics, and in particular the concept of solicitude within it, constitute a response adapted to a particular context and not a set of moral precepts that can be mobilized at will, as Fabienne Brugière points out (see quotation below). The notion of solicitude, given pride of place in the 1980s by psychologist Carol Gilligan in *In a Different Voice* [GIL 82], is understood here as "[the] ethical concern situated, rooted in the complexity of the context and based on deliberation, care and preservation of the relationship with others"[15]. On these premises, care ethics appears capable of guiding the concrete practices of contemporary telecare that call for differentiated, even personalized solutions, according to the pathologies (renal failure, heart failure or diabetes) or according to the people suffering, to varying degrees, from the same disease.

> *When it comes to "taking care", the question is no longer (...) so much to consider the place of values, rules, or laws in action by producing moral reasoning, but to consider the best way to conduct oneself in a particular context, with other subjects carrying social or cultural beliefs, one's own affective histories, etc., as well as the best way to behave. Ethics is never a totally rational matter with concern and care, it invests itself in the sphere of the needs of others and the need for appropriate response, attention to others, maintaining relationships against any hasty condemnation of conduct or any*

14 These guidelines aim, in particular, to encourage practitioners to inform their patients about the limits of new care technologies and to set up personalized patient monitoring. "AMA adopts new guidance for ethical practice in telemedicine", https://www.ama-assn.org/ama-adopts-new-guidance-ethical-practice-telemedicine, June 13, 2016.

15 Source: https://fr.wikipedia.org/wiki/Éthique_de_la_sollicitude.

> *attitude of indifference. Care theory is part of this redefinition of the relationship between ethics and morality.* [BRU 11a, p. 36]

As we shall see[16], care ethics, while respecting medical ethics, can guide contemporary practices of person-centred telecare, a term whose meaning must now be defined.

## 2.3. The person in telecare

Simply put, the person in the telecare context, to use Xavier Guchet's helpful expression, is "the concrete patient who must be put back at the heart of care systems" [GUC 16, p. 208]. The person, as a concrete patient or personal being, is however neither a given nor an asset but rather, as Emmanuel Mounier points out in *Le Personnalisme*, a lived activity of self-creation:

> *The person is not the most wonderful object in the world, an object that we would know from the outside, like others. It is the only reality that we know and that we make at the same time from within. Present everywhere, it is given nowhere (...) It is a lived activity of self-creation, communication, and adherence, which grasps itself and knows itself in its act, as a movement of personalization. To this experience, no one can be conditioned or forced.* [MOU 49, p. 10]

Unlike the concept of individual – in the common sense of biological individual – or individuality – in the sense of individuality defined at the level of the organism and the molecular mechanisms of biological processes – the person, in Emmanuel Mounier's thought, "while preserving their concrete attachments to the world and to life (...) is more than life" [GUC 16, p. 192]:

> *Personality was expected to begin by defining the person. But we define only objects external to humankind, and that we can place under the glance. But the person is not an object. It is even that which in every human cannot be treated as an object. Here is my neighbor who has in his body a singular feeling that I cannot feel; but I can look at this body from the outside, examine its moods, heredities, form, diseases, and in short, treat it as a matter of physiological, medical knowledge.* [MOU 49, p. 9]

In other words, the person can only be understood, on the one hand, as "creation of unpredictable novelty" – a definition that applies perfectly to the person with a

16 See Chapter 8.

chronic disease whose health status changes[17] are not always predictable and, on the other hand, as an individual who escapes control efforts, as Xavier Guchet points out again:

> *To sum up, there is a fundamental difference between the individual and the person: the individual is caught up in a network of social and political interventions, they are obliged to conform to certain norms of behavior; the person on the contrary is what, in the individual, escapes all the efforts of classification and control, it is creation of unpredictable novelty. In the personalist perspective, the person is not a given, an object that one could comprehend from the outside: the person is to be made, they are a self-creation.* [GUC 16, pp. 190–191]

The meaning given to the person, at the center of a relational ethic of telecare, is thus threefold: it refers to the trying subject, to the individual co-responsible for care and to the being who reinvents themself (here under the influence of chronic illness). This definition of the person draws attention to the personal being who can no longer be ignored in the care relationship, at the risk of transforming telemedicine into a dehumanized and dehumanizing *medical service*. Health professionals are now made aware of this, in particular, because of the efforts undertaken by patients themselves, but also because of the development of therapeutic education and new health training programs: practitioners are thus encouraged to *take note* of the personal dimension of patients. However, patient management at this level, in telecare protocols, is far from certain to date.

Faced with the growing demand for personalized care, particularly in telemedical settings, the evidence-based medicine approach alone is too narrow. Defined as "the conscientious, explicit, and judicious use of the best available evidence in making decisions about the care of each patient, (...) a practice of integrating each clinical expertise with the best external clinical evidence from systematic research", [SAC 96], the evidence-based medicine approach, which certainly provides for the compassionate consideration of patients' rights, preferences and experience in treatment decisions, is based on a process that largely involves the clinician's expertise and the best external clinical data (see Box 2.2). These provide scientific evidence and take into account the statistical results of clinical trials.

17 In the Ottawa Charter of the World Health Organization (WHO), health, in a broad definition, means anything that contributes to the physical, moral or social well-being of people. Thus, health mobilizes all areas of human activity. See [WOR 86].

**Steps in the evidence-based medicine (EBM) process**

1) The formulation of a clear and precise clinical question based on a clinical problem posed.
2) Searching the literature for relevant and appropriate clinical articles on the problem.
3) Critical evaluation of the validity and usefulness of the results found ("level of evidence").
4) The application of evaluation results in clinical practice for personalized management of each patient.

**Box 2.2.** *The EBM process*[18]

However, in a recent book, Bruno Salgues identifies the four harmful consequences of EBM (see Box 2.3).

1) "[…] the confusion between risk factors, data and the disease, which is certainly the most insidious and veiled";
2) "[…] a rapid shift from the correlations found through studies to actual causal links. The result is that medicine will treat patients who are not ill and, conversely, that doctors can overlook patients in need";
3) "[…] the success of a drug-based and technical solution is favored. This can potentially entail overmedication, overtreatment and unnecessary operations, all actions that healthcare professionals perform in good faith";
4) "[…] this approach overestimates figures and underestimates clinical observation".

**Box 2.3.** *Consequences of EBM [SAL 16, p. 18]*

On the strength of this observation, Bruno Salgues rightly concludes that "the industrialization of health will have to involve EBM, but under no circumstances will this be the only basis it will rely on" [SAL 16, p. 18]. Telemedicine, a key device in the industrialization of healthcare, is already becoming and will become more independent in the future from EBM.

Radically different from EBM, i.e. biomedical medicine focused mainly on treatment based on clinical statistics, the "person-centered care" approach, used in telemedicine, aims mainly to integrate the (tele)patient as a person in the medical decision-making process and to promote their autonomy in the remote care protocol. Unlike EBM, the "person-centered care" approach is based on a holistic approach to care that takes into account the person in all their diversity: biological, medical,

18 Source: https://fr.wikipedia.org/wiki/Médecine_fondée_sur_les_faits.

social and psychological. It thus integrates factors as diverse as the capacities of the person, the resources at their disposal, the wishes they express, their health, their well-being, without forgetting their sociocultural and relational environment.

One might thus be led to believe that "patient-centered care" or "person-centered medicine" approaches are mainly aimed at deploying care, prioritizing the patient in the care mechanism in a context of advanced medical technology. However, it is not that simple. While it is certainly necessary to recognize that these approaches have a humanist dimension, they jointly aim to refound medical theory or practices as a whole, a reformist trend that, as Xavier Guchet points out, poses theoretical and methodological problems:

> *The "person-centered medicine," "patient-centered care," and "personal medicine" approaches do not intend to articulate science and humanity in a purely external way, by conferring an "extra soul" on a scientific medicine that remains unchanged in its epistemic foundations: on the contrary, it is a question of proposing a new epistemic framework for medical theory and practice, which does not go without posing both theoretical (to conceptually unify heterogeneous medical practices) and methodological (to define the relevant evaluation criteria for patient-centred medicine) problems.* [GUC 16, p. 207]

"Person-centered medicine" approaches generally share with personalized medicine (which strictly means "the establishment of therapeutic protocols based on the genetic characteristics of patients") [GUC 16, p. 49] the same "transforming ambition" by making the person not a concrete being, at the heart of care, but a quasi-abstraction:

> *These alternative approaches [person-centered type] seem to have the same transformative ambitions, both epistemic and organizational, as personalized medicine. In all cases, it is a question of in-depth health system reforms, based on the same type of arguments (inefficiency, costs in particular) (...) Far from being solely the concrete patient that must be put back at the heart of healthcare systems, the "person" now appears as the a priori of any possible theoretical proposal on medicine today, the only conceivable frame of reference within which epistemic and axiological conflicts can be expressed.* [GUC 16, p. 208]

The relational ethics of telecare, described in this book, is inspired by "person-centered" approaches, taking into account the interests of the *concrete patient* and understanding, on the basis of case studies, the need to reconcile today's human values of care with the innovative practices of remote monitoring and care managed autonomously by the patient. This ethic, which borrows jointly from medical ethics and care ethics, questions how to ensure respect for the fundamental right to dignified and quality care, integrating technological and medical developments in a context of ideological overuse, scientific competition and economic or even more financial pressure. As Emmanuel Hirsch points out, "current circumstances, not always reconcilable with ethical requirements, call for a duty of vigilance and rules of governance that should be thought of together"[19]. Beyond this general or political framework, the ethical approach proposed in this book *ultimately* questions the challenge of the human being at the heart of current telecare practices, the person in the test of remote surveillance. This ethical questioning aims mainly to identify an area of research that is still largely undervalued today: the experience of the person in their triple relationship to disease, connected devices and telecare participants.

19 [HIR 10, p. 15]. The author also adds the following comment clarifying our point: "It is also, appropriate to grasp the meanings of care. This notion of care now at the heart of politics confers a meaning other than strictly medical to the act of care. The sick person claims the right to be recognized in their own values and autonomy, beyond what would be the only treatment (cure) of their illness. At the heart of care, the delicate circumstances of the announcement of the illness, those of the shared decision, or even of the refusal of treatment, encourage us to think in other words of the "care relationship"" [HIR 10, p. 17].

PART 2

# Telecare Phenomenology

# Introduction to Part 2

The knowledge acquired through large-scale medical-economic evaluations and feedback on patients' quality of life in France over the past 10 years has made it possible to draw up a twofold observation from which a phenomenological reflection on the person being telemonitored can be formulated: on the one hand, telemonitoring techniques seem to provide a rational vision of the sick body; on the other hand, the management of the experiential and psycho-emotional dimensions of the patient with a chronic disease, at this stage in the evolution of telecare in France, remains confidential. Psychological support is offered to the patient in certain health establishments, but this is still only part of limited local or regional initiatives.

The phenomenological analysis in Part 2 of this book aims to show how the experience of the chronic patient is developed, in a relationship to oneself determined by technique, within the framework of remote medical monitoring and home care protocol. To this end, we examine, on the one hand, the "patient experience" as reflected in the evaluations and feedback documented in this book, as well as normative care practices. On the other hand, we examine the specific physical and psychological disease experience and their temporality, as well as the conditions of care of the person affected within the framework of telecare protocols.

DEFINITION.– Phenomenological analysis refers to the study of phenomena and life experience; in other words, the facts of the physical, psychological or social world.

The phenomenological analysis of evaluations and experience feedback aims here, on the one hand, to show how the patient constructs themself as a subject through contemporary telecare techniques, practices and procedures and, on the other hand, to identify the conditions for the possibility of extended patient management according to their life experience and emotional states.

QUESTION.– What benefits can patients derive from the recognition of their psycho-emotional states within remote monitoring protocols?

To answer this question, we first examine the limitations of a reductionist conception of the body as a rational data set, inherited from the biomedical model (and underlying telecare), and we then shed light on the subjective scope of the patient's confidential relationship to illness and caregivers. Between these two extremes, a third path is also explored: the balance of objective and subjective knowledge about the disease within the framework of the telecare relationship that underlies a new conception of the person.

In short, the objectives of Part 2 is to analyze the criteria used in national and regional assessments of the quality of life of patients with chronic diseases and to propose a phenomenological and epistemological analysis of the temporal and emotional factors related to the disease and its management in the telecare of people with chronic diseases (renal failure, heart failure and diabetes). The results of this study shed light on the conditions for improving the quality of life management of the persons interviewed and, furthermore, pave the way for an ethical framework for telecare.

# 3

# A Cross-dimensional Look at the "Patient Experience"

*It is always the relationship to the sick individual, through the clinic, that justifies the pathological qualification. While acknowledging the importance of objective methods of observation and analysis in pathology, it does not seem possible to speak, in any logical correction, of "objective pathology" [...] the [objective] intention of the pathologist is not that their subject is a matter emptied of subjectivity.*

Georges Canguilhem
[CAN 10, pp. 156–157]

Major medical–economic evaluations of contemporary telemedicine (*HAS*, *CNOM* and *OPIIEC*), feedback on pioneering home medical telemonitoring programs (*Diabéo*[1], *Calydial* and *SCAD*) and the results of surveys conducted among patients suffering from chronic diseases such as diabetes, renal failure and heart failure have made it possible to identify the gains of telemedicine in terms of organization and quality of care. However, a review of this abundant literature leads to another, more mixed observation. The management of the psychological and emotional experience of the disease in the context of remote monitoring, which constitutes the "patient experience", is insufficiently developed and has not yet been the subject of any in-depth clinical evaluation.

1 See: https://www.diabeo.fr/ for more information.

DEFINITION 3.1.– What is meant by "patient experience" in today's telemonitoring environment? This personal neologism of the partial article (experience "of the" patient) here refers to the patient's relationship to telehealth objects and system, and designates their experience of telemonitoring practices. Beyond this medicotechnical dimension, however, the *patient experience* covers another reality: the psychological experience of the disease, the remote monitoring protocol and the caregiver–patient relationship – and their respective temporalities – without forgetting the experience of the emotional states linked to this experience. This chapter aims to capture the specificities of the subjective dimension of the *patient experience*, thus defined, in order to broaden its meaning.

Telecare technologies target disease progression as a priority, while wellness, although targeted through the use of these technologies, is insufficiently taken into account. Its evaluation is often based on general criteria that make it possible to identify the long-term medico-economic viability of remote home monitoring. New telemonitoring technologies now tend to facilitate closer contact between patients and their doctors, greater commitment on the part of individuals in the care relationship and an overall improvement in their quality of life. Telesurveillance, as we have seen, refers to remote monitoring through the analysis of real life data transmitted to the doctor by the patient through connected devices. Telemonitoring differs from remote surveillance, a mode of real-time patient observation, or teleconsultation (real-time video consultation of the patient), and, more generally, from the other telemedicine procedures that can be performed in France, as mentioned above.

DEFINITION 3.2.– The patient's *normative construction* designates the constitution of knowledge of the patient's state of health based on biological standards. This process takes place through statistical measurement technologies that produce "objective" disease knowledge.

This chapter begins by identifying the benefits of telemontitoring patients with chronic diseases in the light of the above-mentioned major medico-economic evaluations and experience feedback. In particular, the examination of the criteria used in these evaluations to measure the *patient experience* makes it possible to understand the way in which the *normative construction* of the remotely monitored patient is currently carried out in a relationship to oneself determined by the technique. This construction is based mainly on the acquisition and interpretation of standardized knowledge about the disease based on *objective* technologies: computerized care devices and communication devices for remote surveillance. The joint examination of the criteria and modalities of protocolized remote care shows, in a second stage, that the disease remains considered from a cause-and-effect perspective, while the psychological experience of the patient – i.e. their intimate temporality – and their emotional states, which play an essential role in the awareness of the disease, seem to be relegated to second place in the hierarchy of knowledge necessary for diagnosis, care and the caregiver–patient relationship.

The proposed analysis in this chapter is twofold. On the one hand, it aims to identify the specificities of the subjective appropriation process of the disease jointly, not competitively, with the objectification of the patient experience that takes place through the use of telecare technologies. In the inviolable interpretative space of the patient lies personal freedom – put to the test of care technicization – to act and to decide what is good for them. Care technicization is understood here as the set of technological means made available to those involved in telecare – both those receiving care and those providing care – to manage chronic disease. Care technicization differs from a broader process known as "health systems industrialization", the mechanisms of which Bruno Salgues analyzed and proposed the following definition:

> *Industrialization is the process of transformation of the production system of a society through technology* [SAL 16, p. 23].

On the other hand, the subjective perception of the disease is understood here, with regard to its objective determination, in the light of the phenomenological heritage of Maurice Merleau-Ponty and Gilbert Simondon, who invite us to requalify the *patient experience* from the prism of the values of temporality and emotional sensitivity. The reexamination of these fundamental values, whose complementarity echoes the inseparability of human *temporal perception* and *affects* demonstrated by the cognitive sciences[2], allows us, because of the phenomenological analysis of perception, to renew the meaning given by telecare participants, to the patient's feeling.

This insight is relevant not to affirm the need to free the patient from the technical norms of care (which would amount to freeing them from the telecare to which they have committed themself), but to identify the conditions for the possibility of a broader management of the patient's *temporal* and *emotional* knowledge through which their *experience* can be built, as well as a *coherent* interhuman therapeutic alliance, in the sense that the philosopher Hang-Georg Gadamer, cited here by the philosopher Philippe Barrier, confers on this term:

> *If a link is thus established between what the patient has felt and what they have been taught, by an explanation that does not oppose subjective feeling and objective determination but puts them in*

---

2 The link between temporality and affects, particularly in relation to the patient's capacity to anticipate, is the subject of a specific examination in Chapter 7.

> *coherence (since "there is coincidence between the medical judgment and the patient's self-awareness" as Hans-Georg Gadamer says), then the disease may appear to them as already actually constituting their experience* [GAD 93, p. 62; BAR 10a, p. 165].

In addition to an inventory of the proven benefits of remote medical monitoring at home, this chapter therefore offers a threefold perspective on (1) the construction of the patient experience and the concept of the person it presupposes in the context of telecare; (2) the review of current remote monitoring practices and their reductionist tendency; and (3) the order of the remotely monitored person whose temporality and sensitivity constitute the very heart of the experience of the disease and its autonomous management at home.

## 3.1. Medico-economic evaluations and feedback

The medical reality of a chronic disease is: permanent symptoms, regular medical and health care interventions (scheduled and unscheduled), a frequent need for hospitalization (frequent complications), permanent treatment and care (no cure), funding for numerous procedures, and an impaired quality of life and independence. Telemedicine and, more specifically, home medical telemonitoring, allow for better management of all these parameters.

Several advantages are recognized today in telemedicine. For healthcare professionals, telemedicine makes it possible first and foremost to reduce geographical isolation and to secure practices. For the health system, it avoids certain costly hospitalizations. The benefits of telemedicine are measured in *time savings* (reduced transport, faster diagnosis, reorganization of working time, shorter hospital stays, shorter waiting times) and *safety* (reduced risks, continuity of care, quality for all, better information and increased medical skills, reduced professional isolation). Added to this list are less testing and/or less redundant tests, fewer orders for additional tests, and a better response to the needs of underserved populations. For (chronic) patients, telemedicine makes it possible to jointly improve access along with the organization and quality of care, as shown in Table 3.1, taken from the study on the information and communications technologies (ICT) health sector in France, conducted in 2012 by OPIIEC.

| Benefits of telemedicine for | patients | doctors | institutions |
|---|---|---|---|
| **Improving the quality of care** | | | |
| Quality of care | • | | |
| Improved diagnosis | • | • | |
| Mortality reduction | • | | |

| | | | |
|---|---|---|---|
| Number of years of life gained (weighted or unweighted by quality) | • | | |
| **Improving the organization of care** | | | |
| Reduction in the number of hospitalizations, reduction in the length of stay | • | • | • |
| Reduction of redundant procedures or tests | | | • |
| Increase in the skills and cooperation of professionals | | • | • |
| Increase in productivity (fewer lost work days) | • | | • |
| Decrease in transport | • | • | |
| Time saving (including reduced wait times) | • | • | |
| **Other benefits** | | | |
| Improving access to care | • | | • |
| Decrease in isolation | • | | |
| Public security | | • | • |
| Access to information | • | | • |
| Regional equality | | | • |

**Table 3.1.** *Main benefits of telemedicine according to OPIIEC*[3]

The development of telemedicine and telemonitoring, nevertheless, faces many obstacles, most of which have been identified in the European Commission's 2014–2020 action plans for health[4]. These action plans have four main objectives: (1) to promote health, prevent disease, and create the conditions for healthy lifestyles, taking into account the principle of integrating health issues into all policies; (2) to protect EU citizens from serious cross-border health threats; (3) to contribute to innovative, efficient, and sustainable health systems; and (4) to improve access to better and safer health care for EU citizens. These plans identify in particular the lack of an interporerability framework, the lack of knowledge of what telemedicine and digital health solutions can bring, the confidence of healthcare professionals, evidence of efficiency, and also report legal and fragmented reimbursement frameworks, often insufficient or non-existent, high investment costs (including for (ARS) [Agence régionale de santé, French Regional Health Agency], uncontrolled judicial or legal aspects – although France has been at the forefront of this field since the 2009 HPST law – and the absence of economic models. These obstacles to the deployment of telemedicine are not all of equal importance and must be

3 *Observatoire paritaire des métiers de l'informatique, de l'ingénierie, des études et du conseil* (OPIIEC) [French joint observatory of IT, engineering, studies and consulting professions]. Study on information technologies for new healthcare organizations. Creating value from the health ICT sector in France, 2012, p. 120.
4 See: http://ec.europa.eu/health/programme/policy/index_fr.htm.

considered differently according to the different acts of telemedicine. In this context, remote monitoring, based on new and complex organizations, is identified, with coordination tools, as a major lever in optimizing the management of patients with chronic pathologies. Nevertheless, currently published evaluations remain divergent on the medico-economic efficiency of such projects. This situation does not (yet) allow decision-makers to deploy on a large scale and set up sustainable financial frameworks, and thus create the conditions for a market for industrialists. One point, however, seems to be reflected in these evaluations: the measurement of efficiency will increasingly depend on the consideration of patient-centered qualitative criteria.

In its April 2012 summary report, *SYNTEC numérique*, one of the French professional trade unions in the digital industry and member of the SYNTEC federation[5], also insisted on the fact that the sustainability of telemedicine will undoubtedly depend more and more on the participation of patients in the financing of telemonitoring services: "the deployment of telemedicine in France was first prompted by health professionals, mainly in the hospital sphere. This led to the first projects being oriented toward medical uses (tele-expertise, remote interpretation). From now on, and this seems obvious for remote monitoring but also for remote consultation, telemedicine must, in order to develop, seek the support of patients [...] If joint investments by public authorities and industrialists are necessary to initiate the development of telemedicine, the sustainability of the sector depends on informing and training patients, through reflection based on their uses of telemedicine, and even ultimately through their participation in the financing of telemedicine services. This strong support from patients would establish a strong growth driver for the development of telemedicine, and would attest to the maturity of the market" [TEL 12].

## 3.2. Evaluation is the key to deployment

To be integrated into the health care system, both organizationally and financially, telemonitoring – and with it the telecare it is one of the mechanisms of – must be evaluated and demonstrated to be efficient. Nevertheless, while there is unanimity on the need for evaluation, there is no consensus on the method and objectives, which partly explains the diversity of available project evaluation results and their sometimes contradictory results. Different methods are proposed to evaluate telemedicine, telemonitoring and telecare today (see Table 3.2): such as, among others, MAST in Europe, a telemedicine evaluation framework published by

5 Source: www.syntec.fr.

the HAS in France, the GEMSA grid[6] (Goff – Pronost, Telecom Bretagne) or the TEMSED model (Andrology, Gerontology, Inflammation, Modeling laboratory).

The TEMSED model is intended to structure technology assessments for autonomy and associated solutions. Developed by Vincent Rialle and his team (AGIM laboratory – CNRS, *Université Joseph Fourier Grenoble*, *École Pratique des Hautes Études, Université Pierre Mendès*, France), the model proposes to evaluate solutions and technologies according to several dimensions (technological, ergonomic, medical, social, economic and ethical). The method is divided into four chronological phases. Transversally, and throughout the evaluation process, an ethical and deontological study is carried out: technical quality study (phase 1), ergonomics study (phase 2), the study of medical and social practices (phase 3), economic study (phase 4). The advantages of the TEMSED solution are the following: the evaluation of projects according to a multidisciplinary approach, open and adaptable collection tools according to the typology of the project, an *à la carte* method with modular dimensions and tools according to the objectives of the evaluation, and a predominant place in the method given to ethics[7].

| Method | Developer | Characteristics |
|---|---|---|
| MAST | European Union | To develop a methodology allowing a reproducible scientific evaluation of the various telemedicine deployment experiments. |
| GEMSA grid | Telecom Bretagne | Telecom Bretagne and the CGIET, *Conseil général de l'industrie, de l'énergie et des technologies* [General council of industry, energy and technology] proposed an evaluation tool for ICT projects in health and autonomy: GEMSA, as part of funding from the Fondation Telecom during the "health ecosystem: simulation and evaluation" project. This evaluation grid should make it possible to: improve and harmonize calls for projects from contractors on related themes with, as a result, commercial gains for companies and greater equality of access to new healthcare offers in the various territories; taking into account the points of view of all |

6 Multidisciplinary evaluation grid for evaluating health projects based on ICT.
7 Source: http://www.gerontechnologies.com/article/technologie-ergonomie-medical-social-economique-et-deontologique. Page consulted on September 25, 2015.

| | | |
|---|---|---|
| | | stakeholders, whatever their nature; highlighting by providers the value of their solutions, whatever their nature; feedback and sharing of experience at all levels; placing projects under control in relation to the initial strategic objectives[8]. |
| HAS Grid | France | Impact matrix of the expected effects of telemedicine. |
| TEMSED Model | AGIR, *Université Joseph Fournier* (Grenoble) | Involves six areas of value (technological, ergonomic, medical, social, economic, ethical, etc.). |

**Table 3.2.** *Four main evaluation grids for telemedicine in France*

These methods have many points in common (multiaxis approaches, taking into account organizations) but also sometimes different methodological choices (method for enhancing patients' quality of life, for example). Beyond these differences, it is essential to be able to carry out large-scale studies (multicenter evaluation, sometimes on an international scale). It is only on the basis of validated and consensual evaluations that these organizations can be put to use by healthcare professionals and patients, that financial models can be put in place and that an industrial and sustainable market can emerge. However, the evaluation must not hinder deployments: the consolidated medico-economic evaluation, taking into account the impact on the organization of care, must be able to be carried out in parallel with the deployment of solutions.

In France, the most comprehensive medical and economic evaluation report to date (286 article references compiled) is that of the *Haute autorité de santé* (HAS), published in 2013. In particular, it proposes an impact matrix of the expected effects of telemedicine in order to "delimit the scope of the medico-economic evaluation by targeting the criteria that can be used to put into perspective the consequences of an intervention in terms of health gains (medical effectiveness) and resources consumed (costs)" [HAU 13, pp. 112–113]. The matrix proposed by the HAS aims in particular to "identify a plurality of dimensions and take into account indicators including the satisfaction of different categories of stakeholders, the quality of life of patients and their families, the effects on the organization and safety of care, social and ethical aspects" [HAU 13, pp. 112–113]. Figure 3.1, extracted from this evaluation report, presents an enlightening reference frame of values (expected and required) of telemedicine.

8 Source: http://recherche.telecom-bretagne.eu/gemsa/. Page consulted on September 23, 2015.

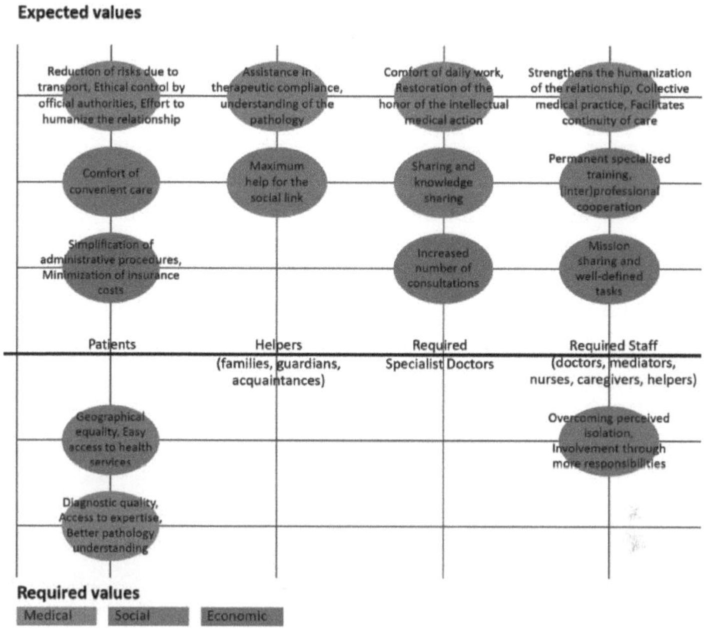


**Figure 3.1.** *Values according to operational subsystem stakeholders [HAU 13, pp. 112–113]. For a color version of this figure, see www.iste.co.uk/bardy/telemedicine.zip*

According to the HAS, as indicated in this reference framework, telemedicine must therefore move in the direction of increased patient care, in terms of human support for the care relationship and to strengthen the ethical control of telemedical practices. However, this reference framework does not identify any value specific to the patient that could be shared by the various players in remote medical home monitoring, and for which better management would increase the patient's commitment to the remote home monitoring protocol. These criteria are also lacking in the telemedicine evaluation model that is currently the most widespread in Europe.

## 3.3. MAST and RENEWING HeALTH

MAST (*Model of assessment for telemedecine*) is an evaluation grid dedicated to the medico-economic analysis of telemedicine experiments. It is a European project, called RENEWING HeALTH (REgions of Europe WorkINg toGether for HEALTH)[9], which aims to develop a methodology allowing a reproducible scientific evaluation of

9 Source: http://www.renewinghealth.eu/en/.

the various telemedicine deployment experiments. RENEWING HeALTH is a European project funded in part by the European Community under the New Technologies Policy Support Programme. RENEWING HeALTH aims to implement large-scale real-life test platforms to validate and evaluate innovative telemedicine services based on a patient-centered approach and a common rigorous evaluation methodology. Involved in this project is a consortium of nine of the most advanced European regions in terms of implementation of new health technologies. France is not part of this European program, although this method is used in the evaluation of some telemedicine devices. The MAST method does not develop a synchronic approach. All parameters must be analyzed at the same time, without any particular phasing.

**Figure 3.2.** *Areas covered by the MAST evaluation*[10]

The technological aspects of telemedicine and/or the patient's emotions are absent from the MAST method (as they are from the HAS repository). Their specific value is not assessed. The evaluation of technologies only focuses on cost aspects (economic analysis) and uses (ergonomic analysis). Yet the ethical aspect appears in every chapter heading of the analysis. However, to date, no psychometric assessment has been formalized for this field. There are, therefore, no preestablished and directly reusable grids in the MAST model, but adaptable questionnaires that can be used to answer questions from the following seven application areas, including that of the patient, chosen by the developers:

– health problem and application characteristics;

– safety (adverse events);

– clinical efficacy;

– patient perspective;

– economic aspects;

10 http://www.gerontechnologies.com/article/model-assessment-telemedicine. Page consulted on October 1, 2015.

– organizational aspects;

– sociocultural, ethical and legal aspects.

The main advantages of MAST are the following:

– good consideration of patients' opinions according to their characteristics (medical and social);

– the approach adopted on the medico-economic aspects of deployments is precise and well-structured.

Although applied within a European framework limited to the nine participating countries[11], the MAST model is used in the evaluation of certain telemedicine projects in France. The URC Eco, the clinical research unit in health economics of Ile-de-France dependent on the AP-HP, which is the national correspondent of the *Observatoire européen des systèmes et des politiques de santé* [European observatory on health systems and policies][12], uses the MAST method for medico-economic evaluations of telemedicine projects entrusted to it[13]. However, the MAST model is not unanimously accepted in France. To date, therefore, there is no consensual evaluation mechanism shared by all participants in the health and autonomy sector. Such a tool is essential to set objectives and assess the results of the deployment of ICT-based offerings in a consistent and transparent manner. This absence is due to the partial nature of the methodologies used so far. Indeed, evaluation systems are now based on practices in the health sector alone. This results in methodological biases and the failure to take into account the non-strictly medical effects induced by the introduction of ICTs. However, the evaluation of a technological project must integrate technical, economic, clinical and organizational parameters without ruling out ethical dimensions and uses, taking into account patients' experiences, opinions, expectations and values.

## 3.4. Toward a psycho-emotional approach to quality of life

In this context, the evaluation of quality of life is an essential element in the field of chronic diseases, as it is one of the main elements that make it possible to judge the effectiveness and quality of patient management as a whole in a medical context. The World Health Organization defines quality of life as follows (1994):

> *An individual's perception of their position in life in the context of the culture and value system in which they live and in relation to their*

11 Nine countries are currently involved in this programme: Italy, Denmark, Sweden, Norway, Spain, Finland, Greece, Austria and Germany.

12 [KID 10, p. 64]. See also [KID 12].

13 See the URC ECO website: http://www.urc-eco.fr.

> *goals, expectations, standards and concerns. It is a broad ranging concept affected in a complex way by the person's physical health, psychological state, personal beliefs, social relationships and their relationship to the salient features of their environment*[14].

A multidimensional concept encompassing patients' social, medical, cultural, psychological and economic factors, quality of life is as subjective a concept as pain can be. Patients and physicians do not share the same conception of quality of life. For the patient, quality of life is a dynamic concept that evolves according to life events and specific parameters: physical state (autonomy, capacity), somatic sensations (symptoms), psychological state and social status (connections, relationships with the social, friendship and professional environments).

Measuring quality of life in a medical context is generally assessed by scales that are neither indices of disease severity, a measure of physical or intellectual performance, nor an index or measure of symptoms (for example the Hamilton scale, see Box 3.1). Quality of life scales can be generalist (i.e. applicable to all medical fields and allowing comparisons by pathology) or specific (i.e. constructed for a particular pathology). Quality of life scales are based on a questionnaire that is administered by the interviewer or self-administered (i.e. completed by the patient) depending on the case. The development and validation of questionnaires are long but essential steps before any use. These questionnaires result in the calculation of scores that are interpreted based on other studies conducted with the general population or a specific population.

> The Hamilton rating scale for depression (HRSD) is a multiple choice questionnaire that American clinicians can use to measure the severity of a patient's major depression. Max Hamilton, who first published the scale in 1960, revised and evaluated it in 1966, 1967, 1969 and 1980. When it was published, it was called the gold standard, but since then, the test and the conceptual basis have been questioned. Hamilton also indicated that his test could not be used as a diagnostic tool.

**Box 3.1.** *HRSD*[15]

### 3.4.1. *The SF-36 questionnaire*

As its name suggests, the SF-36 questionnaire that has been used in chronic kidney disease (CKD) surveys consists of 36 questions that explore eight areas that

14 From http://www.who.int/healthinfo/survey/whoqol-qualityoflife/en/. Quoted in [BAC 14, p. 8].

15 https://en.wikipedia.org/wiki/Hamilton_Rating_Scale_for_Depression.

evaluate both the mental and physical components of health. These eight areas are as follows:

– physical activity;

– limitations due to physical condition;

– physical pain;

– perceived health;

– vitality;

– life and relationships with others;

– mental health;

– limitations due to the psychological state.

The assessment of the physical health component is reflected by the domains "physical activity", "limitations due to physical condition", "physical pain" and "perceived health". The mental health component is evaluated through scores in the domains "vitality", "life and relationships with others", "mental health" and "limitations due to the psychological state". However, no quality of life assessment scale, including the SF-36 scale, specifically takes into account the patient's temporal and emotional dimension. To date, there is no "formal" rating scale for this dimension of telehealthcare for chronic diseases in the home. It would undoubtedly also be appropriate to extend the evaluation to the analysis of the correlation between several indicators, most of which were underestimated in the studies:

– life expectancy;

– quality of life;

– quality of life in autonomy.

An examination of the feedback from the deployment of telecare pilot projects in France, as proposed in the following, does not make it possible to interpret, for example, the lifespan in relation to the autonomous lifespan. It would, however, be interesting to know if the latter has an influence on the quality of life in general in the context of the remote care of chronic diseases. However, by analyzing the feedback, we can identify the priorities and objectives of home protocols by identifying the importance of temporal and emotional factors in the patient's experience. In particular, it is intended to support the phenomenological analysis of the patient's "temporal profiles" that are constructed in the telecare relationship[16].

---

16 See Chapter 5 of this book, *The Person Standing the Test of Digital Clocks*.

RAND CORPORATION HEALTH

RAND > RAND Health > Surveys and Tools > Medical Outcomes Study >

# Medical Outcomes Study: 36-Item Short Form Survey Instrument

## RAND 36-Item Health Survey 1.0 Questionnaire Items

Unformatted version

| 1. In general, would you say your health is: | |
|---|---|
| Excellent | 1 |
| Very good | 2 |
| Good | 3 |
| Fair | 4 |
| Poor | 5 |

| 2. **Compared to one year ago,** how would your rate your health in general **now**? | |
|---|---|
| Much better now than one year ago | 1 |
| Somewhat better now than one year ago | 2 |
| About the same | 3 |
| Somewhat worse now than one year ago | 4 |
| Much worse now than one year ago | 5 |

The following items are about activities you might do during a typical day. Does **your health now limit you** in these activities? If so, how much?

**(Circle One Number on Each Line)**

| | Yes, Limited a Lot | Yes, Limited a Little | No, Not limited at All |
|---|---|---|---|
| 3. **Vigorous activities**, such as running, lifting heavy objects, participating in strenuous sports | [1] | [2] | [3] |
| 4. **Moderate activities**, such as moving a table, pushing a vacuum cleaner, bowling, or playing golf | [1] | [2] | [3] |
| 5. Lifting or carrying groceries | [1] | [2] | [3] |
| 6. Climbing **several** flights of stairs | [1] | [2] | [3] |
| 7. Climbing **one** flight of stairs | [1] | [2] | [3] |
| 8. Bending, kneeling, or stooping | [1] | [2] | [3] |
| 9. Walking **more than a mile** | [1] | [2] | [3] |
| 10. Walking **several blocks** | [1] | [2] | [3] |
| 11. Walking **one block** | [1] | [2] | [3] |
| 12. Bathing or dressing yourself | [1] | [2] | [3] |

During the **past 4 weeks**, have you had any of the following problems with your work or other regular daily activities **as a result of your physical health**?

**(Circle One Number on Each Line)**

| | Yes | No |
|---|---|---|
| 13. Cut down the amount of time you spent on work or other activities | 1 | 2 |

| | | |
|---|---|---|
| 14. **Accomplished less** than you would like | 1 | 2 |
| 15. Were limited in the **kind** of work or other activities | 1 | 2 |
| 16. Had **difficulty** performing the work or other activities (for example, it took extra effort) | 1 | 2 |

During the **past 4 weeks,** have you had any of the following problems with your work or other regular daily activities **as a result of any emotional problems** (such as feeling depressed or anxious)?

**(Circle One Number on Each Line)**

| | Yes | No |
|---|---|---|
| 17. Cut down the **amount of time** you spent on work or other activities | 1 | 2 |
| 18. **Accomplished less** than you would like | 1 | 2 |
| 19. Didn't do work or other activities as **carefully** as usual | 1 | 2 |

20. During the **past 4 weeks,** to what extent has your physical health or emotional problems interfered with your normal social activities with family, friends, neighbors, or groups?

**(Circle One Number)**

Not at all 1

Slightly 2

Moderately 3

Quite a bit 4

Extremely 5

21. How much **bodily** pain have you had during the **past 4 weeks**?

**(Circle One Number)**

None 1

Very mild 2

Mild 3

Moderate 4

Severe 5

Very severe 6

22. During the **past 4 weeks,** how much did **pain** interfere with your normal work (including both work outside the home and housework)?

**(Circle One Number)**

Not at all 1

A little bit 2

Moderately 3

Quite a bit 4

Extremely 5

These questions are about how you feel and how things have been with you **during the past 4 weeks**. For each question, please give the one answer that comes closest to the way you have been feeling.

How much of the time during the **past 4 weeks** . . .

**(Circle One Number on Each Line)**

| | All of the Time | Most of the Time | A Good Bit of the Time | Some of the Time | A Little of the Time | None of the Time |
|---|---|---|---|---|---|---|
| 23. Did you feel full of pep? | 1 | 2 | 3 | 4 | 5 | 6 |
| 24. Have you been a very nervous person? | 1 | 2 | 3 | 4 | 5 | 6 |
| 25. Have you felt so down in the dumps that nothing could cheer you up? | 1 | 2 | 3 | 4 | 5 | 6 |

| | | | | | | |
|---|---|---|---|---|---|---|
| 26. Have you felt calm and peaceful? | 1 | 2 | 3 | 4 | 5 | 6 |
| 27. Did you have a lot of energy? | 1 | 2 | 3 | 4 | 5 | 6 |
| 28. Have you felt downhearted and blue? | 1 | 2 | 3 | 4 | 5 | 6 |
| 29. Did you feel worn out? | 1 | 2 | 3 | 4 | 5 | 6 |
| 30. Have you been a happy person? | 1 | 2 | 3 | 4 | 5 | 6 |
| 31. Did you feel tired? | 1 | 2 | 3 | 4 | 5 | 6 |

32. During the **past 4 weeks,** how much of the time has your **physical health or emotional problems** interfered with your social activities (like visiting with friends, relatives, etc.)?

**(Circle One Number)**

All of the time 1

Most of the time 2

Some of the time 3

A little of the time 4

None of the time 5

How TRUE or FALSE is each of the following statements for you.

**(Circle One Number on Each Line)**

| | Definitely True | Mostly True | Don't Know | Mostly False | Definitely False |
|---|---|---|---|---|---|
| 33. I seem to get sick a little easier than other people | 1 | 2 | 3 | 4 | 5 |
| 34. I am as healthy as anybody I know | 1 | 2 | 3 | 4 | 5 |

| | | | | | |
|---|---|---|---|---|---|
| 35. I expect my health to get worse | 1 | 2 | 3 | 4 | 5 |
| 36. My health is excellent | 1 | 2 | 3 | 4 | 5 |

**ABOUT**

The RAND Corporation is a research organization that develops solutions to public policy challenges to help make communities throughout the world safer and more secure, healthier and more prosperous. RAND is nonprofit, nonpartisan, and committed to the public interest.

1776 Main Street
Santa Monica, California 90401-3208



**Figure 3.3.** *Generalist questionnaire SF 36*[17]*. Available at: https://www.rand.org/health/surveys_tools/mos/36-item-short-form/survey-instrument.html*

17 [WAD 92]: https://www.rand.org/health/surveys_tools/ mos/36-item-short-form/survey-instrument.html.

## 3.5. Techniques and self-experience

QUESTION 3.1.– How is the chronic patient experience defined today in the context of telecare?

The analysis of the main criteria used in the evaluations and feedback as well as the synthesis of the medico-economic, therapeutic and psycho-social benefits of remote monitoring can provide valuable elements of response that are essentially medical. Quality of life is certainly well taken into account, but the criteria used to measure this seem to be limited to the material and/or therapeutic benefits of (tele)care. According to these evaluations, *patient experience* is built mainly in the person's triple relationship to the medical device (defibrillator, cycler, insulin pump), disease and therapeutic treatments. As medical and economic studies show, the benefits in these areas are clearly identified, as are the margins for progress and future prospects. Telecare thus seems to construct a certain *concept of the person* at the knowledge crossroads, resulting from the evaluations, the practices it implements (telemonitoring, self-managed care), and the official objectives it sets for itself.

DEFINITION 3.3.– The "objective self" refers to the individual identity that is constructed from objective knowledge, i.e. standardized, general and abstract knowledge distinct from "situated knowledge", i.e. concrete and local knowledge.

More precisely, this concept of a person is that of an objective self, according to the expression of the American anthropologist, Joseph Dumit, i.e. an identity construction based on "the physical and mental attitudes that we adopt according to scientific and medical facts" [DUM 04, p. 7]. This conception of identity differs from the practice known as quantified self, which consists of collecting at regular intervals the real-life physiological data of individuals, such as weight, heart rate and the number of steps taken each day, thanks to devices. Quantified self is a practice[18], a technique of self, where objective self, in the sense defined by Dumit, refers specifically to the *representation of self* constructed from collected and/or averaged life data and medical data.

The meaning of objective self differs from the meaning attributed to the idea of self as individual self-representation. Objective self is an external top-down conception of the person: identity and experience are built from objective knowledge produced here by telecare objects, i.e. medical data collected by devices

18 The authorship of the expression "quantified self" is generally attributed to two journalists of the American magazine, *Wired*: Gary Wolf and Kevin Kelly.

integrated into the patient's daily life. In Dumit's wake, we have seen that these objective data, transmitted to the practitioner and the medical team, shape the construction of the patient. The manufacture of this knowledge, incorporated by the individual, results in preferring a singular mode of representation and conceptualization of the being; the construction of an objective self:

> *The fabric of an objective self is how we appropriate facts about ourselves... facts that we have read, heard, or learned about in the world and incorporate into our own lives* [DUM 04, p. 164].

The making of an objective self, however, is not just about how knowledge, medical or otherwise, makes sense to individuals. It also points to the possibility of knowledge sharing and in this, according to Dumit, carries with it "the hope that responsibility can be multiplied; that the duty of engagement in the making of an objective self can be shared by experts, mediators, and ordinary people" [DUM 04, p. 168]. In other words, self-making does not obscure the participatory dimension of the process of producing and receiving (medical) knowledge. On the contrary: it highlights the recognition of what Donna Haraway calls "situated knowledge":

> *The representation of the object of knowledge [must] be that of an actor or an agent, not that of a screen, a terrain, or a resource; never finally that of a slave subjected to the master who holds the reins of dialectics in their capacity as a singular agent and producer of "objective knowledge"* [HAR 88, p. 592].

The opposition between "situated knowledge" and "objective knowledge" makes little sense today, as Haraway points out, particularly in a context of increased knowledge sharing among participants in the field. This opposition can no longer be justified by the concept of "epistemological rupture", which would require common knowledge (situated), secular knowledge and scientific knowledge (objective) to be totally separated. As Agnès d'Arripe points out, in her critique of situated and objective knowledge: "knowledge is developed in a process of co-construction with participants in the field". This shared knowledge concerns the so-called "expert" and "lay" knowledge, the main characteristics of which d'Arripe recalls:

> *Expert knowledge [is] standardized, general, and abstract knowledge that allows action at a distance; and secular knowledge [is] knowledge made of concrete, situated knowledge, therefore highly diversified and applying to a dense and multidimensional reality* [HAR 88, p. 592].

Within the framework of this double "epistemological" regime – a regime of juxtaposition and not of opposition of knowledge – the patient is situated upstream of expert knowledge (interpretation, diagnosis) constituted from life data transmitted by care devices, which are installed and self-managed in the person's home and/or mobile devices (smartphone and health application, insulin pump, cycler, pacemaker). In this process of constructing the person as an "autonomous agent" of care, producer of data and receiver of "objective" medical knowledge, the patient experience, nevertheless, covers another practical reality: that of the experience of the illness, of the relationship to oneself – constitutive of an intimate knowledge of the illness – and that of the person who is experiencing it, whose psycho-emotional dimension must be taken into account. The review of evaluations and feedback on telecare clearly shows the need to increase patient management at these two levels in tele-monitoring protocols (all pathologies combined). This care is now possible, as evidenced by certain initiatives targeting the psychological support for patients.

Volunteers or others, these initiatives can be conducted outside of telecare protocols, outside a clinical setting, to ensure better management of the patient experience dimension. One example is the *Maison des usagers de Périgueux*, which groups together a multiassociation patient reception and assistance service, or *Élan solidaire*, offered by the *Fédération française des diabétiques*, which is a program to support diabetic patients based on meetings, exchanges and support. Group sessions or individual interviews and these meetings are facilitated by volunteer expert patients. Access is free and open. *Élan solidaire's* goal is to break the loneliness caused by diabetes[19].

These initiatives tend to prove, on the one hand, that the concrete organization of the psycho-emotional care of (tele)patients is essential today and, on the other hand, that the updating of knowledge sharing (experts and laymen), placed here in their telemedical context, calls for the multiplication of meeting spaces, conducive to reflection, following in this respect the research conducted today around inventive collaborative methods:

> *The construction of scientific knowledge is [...] a permanent coming and going between universal and singular, singular and universal; between secular knowledge and expert knowledge. If there is unanimity within scientific research, there is another element on which there is less consensus: the way in which this link between knowledge from the field and more theoretical knowledge should be thought of. All too often, practitioners' knowledge enriches that of the researcher in the first instance and it is only in the second instance,*

19 Source: www.afd.asso.fr.

> *more or less long and mediated by numerous filters, that theoretical knowledge reflects on practice [...] At the present time, it is essential to reflect on alternatives enabling genuine collaboration and mutual enrichment to take place within the same temporality and the same space [...] The implementation of our research and more particularly the reflection work carried out during the Group Analysis Method [...] had an impact on the concrete, situated, and highly diversified knowledge that we have previously pointed out as characteristics of secular knowledge. Through the conflict of interpretations around the chosen story, participants saw their understanding of the situation become more complex*[20].

The experiential knowledge acquired by the patient in their dual relationship to the disease and to the care devices (connected and/or computerized objects) deserves to be shared more widely by all the care participants. Patients cannot be left to fend for themselves in the face of new health technologies. Faced with the isolation conditions of certain patients who are in need of empathic listening, the integration of this knowledge within the framework of a remote monitoring protocol, often complex, is critical today.

This protocolized remote monitoring cannot mask the fact that the experience of illness and care is *first and foremost* a complex medical process. Thus, after the original trauma, the patient must find internal and external resources to heal and cope with the various dimensions of the chronic disease that requires efforts in terms of management and acceptance. Each dimension comprises a specific space and time, new constraints, support factors, personal means and, in the case of remote monitoring, adapted technological means. These factors can be summarized as follows:

– *the original trauma*: triggers pain, distress and stress (pre- or post-traumatic);

– *the environment*: the patient's home is transformed into a microunit of care (the medical sphere takes over the intimate space);

– *the telemedical dimension*: it includes the treatment, the constants (analyses, examinations performed by the patient), the medical history, and the follow-up of health professionals;

– *the emotional relational dimension*: consisting of the personal history, the possible proximity of the family, and the empathy of the social carers;

20 Quote from [DAR 13, p. 231].

– *the social dimension*: for some patients, the "hospital at home" may be perceived as a social place. The telemonitoring players are virtually brought in and health professionals are received. It is both the place of transformation and the space of new life.

These dimensions of care can also generate various emotional states: confusion, loneliness, anxiety, derealization, amnesia and/or hypermnesia associated with the disease (a heightened recollection of moments in life). In our qualitative survey, respondents acknowledged the initial difficulty of adapting to a biological "new deal" and to technological support for care. The introduction of care devices into everyday life, connected or otherwise, initially disrupts the future of the patient–physician relationship, raises fears of isolation and anxiety, and may help to dematerialize or even derealize the disease that has suddenly become "measurable" or "quantifiable". In this lies the complexity of the patient's experience as a person.

QUESTION 3.2.– Under these conditions, why does remote patient monitoring seem to favor the disease *overall*, to the detriment of the person?

There are psychological and ethical monitoring structures in some hospitals, in particular within the framework of the monitoring of kidney patients, such as the *Commission des relations avec les usagers et la qualité de la prise en charge* (CRUQ, Committee on relations with users and the quality of care) of the Calydial association in Auvergne. We should also mention the excellent therapeutic and psychological support provided, among others, to patients on autonomous dialysis at home, by Calydial and the nephrology department of Tenon Hospital (Paris). But the number, means and scope of action of these structures, according to patients, doctors, and medical institutions themselves, remain generally insufficient to date. This observation raises, among other things, the question of the distribution of the means allocated but also that of the principle of equity with regard to psychological follow-up in the regions.

There are significant regional disparities in terms of the allocations of technical resources (follow-up structures) and human resources (nurses, psychologists). One hospital may be better equipped than another at these two levels. This disparity reinforces the feeling that the technique would prevail over the patient relationship. But on closer examination, the relative "forgetfulness" of the patient is not only due to a (relative) lack of means. The primacy given to the techno-medical dimension of the care relationship is undoubtedly also due to the status granted to the patient, inherited from a tradition of biomedical medicine. The perpetuation of the patient's status as an "object of care", inherited from biomedical medicine, in fact results in keeping at a distance the temporal and emotional factors of the disease, which play a major role in the construction of the self and in one's daily management. Requalifying the person's temporality and sensitivity in relation to their experience is therefore enlightening for two essential reasons: on the one hand, it makes it possible to reassess the epistemological value of the patient's experience and, on the other hand, to question the priority given today, in telecare protocols, to the techno-therapeutic dimension of care. The critical reexamination of the techno-medical primacy in the context of telemonitoring at home, jointly brings out the current challenges of the telemedical institution: to accompany patients with chronic disease as best as possible and to assume a *telemedicine of the person* capable of breaking with a medical discourse of a biological nature.

## 3.6. The telepatient: the neutral object of a symptom to be deciphered?

DEFINITION 3.4.– "Telepatient" refers, in the broadest sense, to any sick person whose care is provided remotely because of new care technologies. This follow-up can take place in the patient's home, as in the case of renal failure, or in mobility, as in the case of carriers of pacemakers or defibrillators, and diabetics using a detachable insulin pump coupled to a smartphone equipped with companion software.

Healing for the (tele)patient with a chronic disease is not suppressing the disease, but living with it. This conduct involves reexamining the wellness objectives targeted by remote surveillance by extending it to the subjective dimension of the person, what Philippe Barrier calls the *ethical care relationship*:

> *[Thanks to the ethical care relationship], it is a question of integrating the objectal dimension of the patient (inherent in the therapeutic relationship) into their subjectivation of the disease, instead of reducing it to a dimension thought of as purely objective, i.e. essentially measurable, quantifiable. This medical reductionism [...] mutilates the very knowledge of the disease of an essential part, at the same time as it deprives the doctor of fundamental support for the implementation of therapy. Indeed, after initially denying the patient's subjectivity in order to make it the neutral object of a symptom to be deciphered, they have recourse, in a prescriptive or even imperative mode, to their will so that they are compliant with the treatment. But above all, this objectivistic reductionism differs the relationship indefinitely. The medical relationship must therefore face particular difficulties in developing itself as an authentic and ethical relationship, that is, as a true care relationship* [BAR 14a, p. 158].

The dangers induced by objectivist reductionism, mentioned earlier, and on the status granted to the diabetic patient vis-à-vis their doctor, are exposed here. It is undoubtedly necessary to qualify this by recalling that the diabetic patient being *telemonitored* is not entirely treated as the "neutral object of a symptom to be deciphered" since they may benefit – in certain cases – from psychological follow-up as part of the caregiver–patient dialogue: this dialogue, when it takes place, allows the patient to voice their needs. The conclusions drawn here, by Barrier, on an integrative medicine sensitive to the temporality and affects of the patient, can however apply to the situation of the diabetic patient who is *telemonitored* remotely. Within telecare protocols, it is indeed important today "to integrate the necessary objectification of the medical view of the patient with a perspective more open to their subjectivation, that is to say to their sensitive, affective, intuitive and sometimes already rationalized experience of the disease. This essential recognition of their quality as a subject, testing, interpreting and expressing, would be a decisive first step toward an authentic care relationship" [BAR 14a, p. 162].

The emotional dimension of the patient is never completely denied in a remote monitoring protocol. It can even be monitored on a regular basis, depending on the health care facility. But the management of the experience and affects aims to relieve the patient and not to fully integrate their intuitive knowledge regarding the disease. Emotional states such as stress and anxiety are relegated to the rank of "metrics" and do not directly "influence" the direction of a telemonitoring protocol. In other words, emotional support to the patient does not significantly influence the diagnosis. A better consideration of the temporal dimension (physical experience of the disease) and the emotional dimension (experience based on affects) would undoubtedly make it possible to better manage certain key parameters of telecare: compliance, treatment modalities, anticipation of alerts, compassionate listening and knowledge of the patient's environment.

Recognizing the patient as a subject implies accepting and taking into account the patient's interpretation of the disease as lay knowledge, but without competition with medical knowledge, in telecare. In this inviolable interpretative space, or in this subjectivity, lies the freedom of the person being monitored to act and decide what is good for them. This inviolable interpretative space jointly expresses the capacity of the subject to mature the terms of their consent and their conformity to the protocol. Beyond the two extremes of the reductive objectification of the patient and the free interpretation of the disease seem in fact to emerge the conditions of the possibility of an integrative telecare relationship open to the patient's choices.

By recognizing and welcoming the patient's freedom of choice and interpretation – which is the case in some protocols – without calling into question the relevance of the measurements made within the framework of a follow-up, telecare can in fact take note of the subject's ability to define for themselves their normal state of health, a patient's prerogative according to Georges Canguilhem:

> *It is true that in medicine the normal state of the human body is the state that one wishes to restore. But is it because it is intended as a good end to be obtained by therapy that we call it normal, or is it because it is considered normal by the person concerned? We think it's the second relationship that's true* [CAN 10, p. 77].

Recognition of the patient constructed health standard is arguably one of the main goals that can be achieved through the broader integration of experience and affect into telecare. Because of this recognition of the patient's subjective norm, correlated with the patient's technical normatization, it becomes possible to push back, if not the limits of the normal and the pathological, then at least the borders of the primacy of the medical interpretation of the pathological reevaluated here, in the wake of Canguilhem, by Barrier:

> *Priority is given to the suffering of the patient, whose medicine, with its scientifically assisted technique, constitutes an attempt to respond. We can therefore say that the term "pathological" does not describe an objective but a subjective category. It is the reality of the suffering experienced that is the first and determining factor in the response to this call that it arouses. Life is not the material that would be the object of medical practice, it is that which medicine rushes to answer the painful call of the one in whom it is failing. This absolute priority, both chronological and essential, of the patient's experience, overshadows the cognitive aim of medicine* [BAR 10, p. 159].

The remotely monitored patient can construct an individual health standard in addition to their protocol, but they can only take full advantage of this knowledge if they can judge the therapeutic effectiveness of the protocol in which they are engaged and for which they have been trained. The issue of therapeutic education is therefore central to the patient and to the success of home care. Therapeutic education in the sense of double learning is to be understood here. On the one hand, it is a question of providing patients with the information and conceptual tools necessary to better understand chronic diseases and its follow-up through intermediary technologies (understanding the treatments proposed based on tangible medical data, such as elevated blood sugar levels, cardiac arrhythmias, and/or renal dysfunction). On the other hand, it is essential that any therapeutic education program be jointly fuelled by the knowledge of the person, called secular knowledge, which are mentioned previously. Without this double dimension, the patient cannot judge the validity of the protocol or take full advantage of their experience. It should be stressed that the technical dimension of care cannot be ignored. Conversely, without the contribution of the patient's subjectivity and an extension of their psychoaffective care, particularly at the stage of therapeutic education, telecare remains devoted to reductionist objectivism.

To conclude, evaluations and feedback on remote monitoring show that the relationships to oneself and to the disease are determined by technical criteria, relegating the subjective dimension of the disease, as well as the perception of the therapeutic effects of the protocol, which constitute the patient's experience and the construction of a subjective standard of self, to the background of the hierarchy of knowledge essential for diagnosis, follow-up and care, but this must be qualified. The situation is not the same according to the pathology, the protocol, the region, etc. Telemedicine participants recognize the need to accompany the sick person psychologically as much as possible. The need to better take into account the temporal factors and emotional aspects of the disease is emerging, as shown by the psychological follow-up initiatives conducted with patients in certain care institutions in France.

Between a strictly reductionist conception of the sick body, based on the analysis of rational data within the ethical framework of a strict telemonitoring protocol, and a still confidential approach to the patient's subjective dimension[21], a third approach, nevertheless, seems to be emerging today: that of an extended patient management in their emotional and psychological dimension taking full advantage of the processes of telecare, whose benefits and flaws can be identified through phenomenological analysis. However, to understand this rallying point of the technicality and humanity of care, it is essential to reexamine the values of the patient's temporal experience and sensitivity. In this respect, the phenomenology of perception and philosophical psychology can provide valuable insights into the articulation of these values in today's telemedical context.

## 3.7. Temporality and sensitivity at the heart of the experience

What are the conditions that must be met to move from a conception of the patient as the "neutral object of a symptom to be deciphered" to a broader point of view on the "care subject", as a *chronosensitive* living being[22], not reducible to its somatic and psychological dimensions alone? As we will see, the subject's temporality and sensitivity constitute both their experience and a self-standard that is

21 Although the person's stress and anxiety are subject to therapeutic follow-up by family and/or social caregivers (caregivers, occupational therapists), the exchanges between these participants, around these emotional states, generally take place under the seal of confidentiality, which explains why the experiential and emotional dimensions of the disease have received, to date, only limited attention in studies and clinical evaluations of remote medical monitoring at home.

22 The "chrono-sensitive" being here designates the individual in their double temporal and psycho-emotional dimension. For a more detailed discussion of the concept of chronosensitivity, see Chapter 8 of this book, "Chrono-sensitivity: From Concept to Ethics".

likely to be better integrated into future practices of *person-centered (tele)medicine.* Remote monitoring is a technological device that is fundamentally centred on the patient: it is made up of interfaces and communicating systems that are intended to make the remote patient an end user. Person-centred ethical approaches in telemedicine, such as those encouraging patient participation in the design of these technologies[23], propose a conceptual basis favoring a *holistic* vision of the person: it is indeed a question of taking into account the mental and physical aptitudes of the patient, but also their beliefs and values, their cultural and socioeconomic framework, and their will to draw lessons from their experience of remote surveillance. They also advocate the use of active listening and attention to the actual needs of the patient[24].

In *Phenomenology of Perception (La phénoménologie de la perception),* Maurice Merleau-Ponty advances the hypothesis that the analysis of time can serve as a starting point for understanding subjectivity, and not the other way around. "Analyzing time" writes Merleau-Ponty, "is not drawing conclusions from a pre-established conception of subjectivity, it is accessing its concrete structure through time" [MER 15, p. 472]. The philosopher's intention is not to understand the subject in its "pure form" but rather to seek it "at the intersection of its dimensions", hence his will to consider time *itself.*

However, the perspective adopted in this chapter is significantly different since the perception of time is situated within a concrete framework – that of telecare – and aims primarily at capturing the value of temporality – which designates *the character of what exists in time (and not time itself)* – in the construction of subjectivity and *patient experience.* Merleau-Ponty's phenomenological legacy makes it possible to explain the "concrete nature" of telepatient subjectivity and its links with the temporality of the disease:

> *We will never understand how a thinking or constituent subject can ask or perceive themself in time. If the I is Kant's transcendental I, we will never understand that it can in any case merge with its aftermath in the intimate sense, nor that the empirical I is still a self. But if the subject is temporality, then self-position ceases to be a contradiction, because it expresses exactly the essence of living time. Time is "self-affection by oneself": the one who affects is time as a push and passage towards a future; the one who is affected is time as a developed series of the present; the affecting and the affected are one, because the push of time is nothing other than the transition from a present to a present. This ekstasis, this projection of an undivided power in a term present in it, is subjectivity* [MER 15, p. 488].

---

23 See [CLE 17].

24 See, for example, [BRA 14].

Merleau-Ponty develops here two arguments that resonate with the experience of the person being monitored. On the one hand, the subject of telecare may not construct an identity "out of time", or that is "transcendental" according to the Kantian terminology used here by Merleau-Ponty, but may perceive the joint acquisition of a temporal knowledge of self and illness ("empirical self") and an objective knowledge of the body ("objective self"). Thus, the telecare patient can *affect*, to quote Merleau-Ponty, their perceptive capacity to this mixed temporality resulting from subjective and objective knowledge of themself. This "affection" is doubly positive: it can modify the consenting patient's open outlook on the knowledge of the disease, acquired through the protocol and constituting their experience, and it can jointly influence the bias of truth underlying the remote monitoring of the "objective body" that is imposed on them. For Merleau-Ponty, the bias of the "objective body" of the individual is not the truth of the "lived body":

> *The objective body is not the truth of the phenomenal body, that is to say the truth of the body as we live it. It is only an impoverished image, and the problem of the relations of the soul and the body does not concern the objective body which has only a conceptual existence, but the phenomenal body. What is only true is that our open and personal existence rests on a first acquired and fixed foundation of existence. But it could not be otherwise, since the dialectic of the acquired and the future is constitutive of time* [MER 15, p. 495].

In all cases, the lived body, or the *phenomenal* body, is the seat of consciousness (what Merleau-Ponty calls the "relations of the soul and the body"); it is also the space of the confrontation of the senses with the process of objectification of the self. Temporality is both the zone where being and consciousness coincide and the space within which the confrontation of knowledge from the senses and objective knowledge of the body can be negotiated. In telecare, this confrontation is part of the *patient experience.* In other words, at the peak times of the telecare protocol (dialysis session, insulin injections, control of biological parameters) and at the objective measurement of the evolution of the disease, the patient's non-quantifiable temporal and sensitive regime is superimposed, but bearing relevant knowledge: experience of the constraints imposed by the disease and the telecare device, time management capacities and adaptation to the alerts transmitted because of the telemonitoring and orientation of the person in time (past, present and future).

The subjective knowledge of the person in the context of telecare is constructed from emotional states that nourish their "sensitivity", which can be defined in two distinct ways: either as the ability to be moved, to experience feelings of humanity for others, or as the ability to perceive through the senses. Extending these two complementary definitions, sensitivity in the context of telecare refers to the patient's ability to feel and manage emotions about the illness and its management at home in a technological environment (care and telemonitoring devices) and to build a unique experience of self, at *home*, that shapes the relationship with illness and others. The sensitivity of the monitored patient is thus dependent on a set of human and technical factors, in line with the definition proposed by Gilbert Simondon:

> *Sensitivity alone is not a complete function. Its exercise and its modalities are linked to the overall activity of the organism inserted into their environment, their needs, their way of life, their motivation, and their system in action* [SIM 15, p. 361].

To understand sensitivity, according to Simondon, we must start from the physiology of sensitivity, which implies looking at the body without neglecting its relationship to the environment:

> *The physiology of sensitivity leads to the study of the organism inserted in their environment, because sensitivity is not closed on itself. It is not itself a complete and independent function. The messages it brings modify the organism, but the play of sensitivity depends on the other functions, and is modified by the state of the organism and by their ongoing behavior* [SIM 15, p. 377].

To analyze the physiology of sensitivity, Simondon uses Sherrington's tripartite classification: "The sensory organs are either *exteroceptive* (touch, smell, vision, hearing) or *interoceptive* (stato-receptors sensitive to changes in the body's position in space; nerve endings related to visceral sensitivity, belonging to the sympathetic system, in the form of nerve threads ending in plexus in the vicinity of the epithelia, inside the connective tissue), or finally *proprioceptive*. Proprioceptive organs collect essentially endogenous stimuli, informing the nervous system about the body's functioning, which allows control, command, and regulation (neuromuscular spindles, carotid sinuses)"[25]. This description borrowed from Sherrington shows that the physiology of sensitivity is a complex process that relies on both the senses and the internal "mechanics" of the body in interaction with an environment[26]. Sensitivity is a stimulus that triggers a response to the environment and not pure perception projecting itself onto the world. Sensitivity is the expression of the nature of the individual, of their essence of being trying, that is, of what constitutes them as what they are and gives them their fundamental reality. In the context of telecare, sensitivity finds *its foundation* in the fact that it is at the origin of the intuitive

---

25 [SIM 15, p. 399]. This description by Simondon is adapted from the following work: [SHE 10].

26 For a more in-depth examination of the relationship between affects and temporality, see Chapter 7 of this book, "Subjectivizing the Future: or, the "Patient–Project" Temporality".

production of an intimate knowledge of oneself and the disease that superimposes itself on the objective body.

Sensitivity provides the basis for the patient's complex interpretation mechanism. It can give a basis to the subjectivation of the disease. In psychoanalytic vocabulary, subjectivation is often presented as a psychic work, particularly marked in adolescence, distinct from the identity reworking of the self. Subjectivation is understood here in the philosophical sense, as the action of subjectivating, of bringing to oneself and one's consciousness. In this book, subjectivation refers to the personal process of explaining the disease to oneself and the resulting construction of oneself[27]. As Simondon points out, however, sensitivity is "not very capable of giving an unequivocal knowledge of objects or a coherent and universal appreciation of cases: its data cannot be realized, become certain clues to a particular isolated reality or criteria of a defined situation" [SIM 15, p. 362]. However, sensitivity can distinguish the essential from the accessory. It is thus doubly essential if one follows Simondon in the footsteps of Merleau-Ponty: on the one hand, it is inscribed in the essence of being and finds in it the foundation of its recognition, or more precisely of a possible recognition of the ontological and epistemological value of the essences that underlie it. The concrete translation of this recognition into a telemonitoring context can involve a reinforcement of the patient's *active listening*, making it possible to identify their *essential* character as a person, but it also takes place by a reevaluation of the medical institution of the patient's knowledge resulting from the more or less intuitive interpretation of their emotions, positive (feeling of comfort conferred by the protocol) and negative (stress and anxiety linked to the fear of alerts or health complications), induced by their intimate relationship to telecare devices. Active listening is a concept developed from the work of American psychologist Carl Rogers. The concept is also referred to as that of a *benevolent listener*. Initially designed to accompany the expression of emotions, it is operational in face-to-face situations where the professional *actively listens* to the other. It consists of putting into words the emotions and feelings expressed tacitly or implicitly by the interlocutor. Active listening is finer than reformulation in that it is not limited to saying otherwise what a person has just expressed, but to decoding the generally nonverbalized affective dimension[28].

It is possible to see in this broad recognition of the sensitivity and temporality of the patient one of the major issues, of Kantian inspiration, of telemedicine to come. The second formulation of the categorical imperative already condemns the instrumentalization of the human being (considering it as a means and forgetting it

27 Its ethical evaluation (see Chapter 8 of this book) borrows from the history of modes of subjectivation proposed by Michel Foucault in *L'usage des plaisirs (Histoire de la sexualité II)*.
28 Source: https://fr.wikipedia.org/wiki/Écoute_active. Page consulted on November 3, 2015.

as an "end in itself") and establishes unconditional respect for human dignity. Recognizing the patient in their temporality and their sensitivity to the suffering of chronic illness and in their potentiality to be themself therefore undoubtedly represents, in the Kantian wake, one of the current means of showing concrete respect.

DEFINITION 3.5.– The imperative is a first person command. It is about telling myself what I need to do. It is about deciding how I should act. With Kant, we distinguish two types of imperative. The first is hypothetical. Our action in this context is conditioned by a goal that we have set for ourselves. Interest comes into play. It is about efficiency. The second imperative is categorical, namely that it bears no condition, except to be the product of reason. For example, I must tell the truth at every opportunity and at every moment. The categorical imperative therefore eliminates any particular purpose. It refers to the universal: all of humankind must act in this way. It is a standardization of individual conduct. The categorical imperative is certainly necessary, but also dangerous. Necessary for morality, which dictates, among other things, respect for others. Dangerous because it grows to aboutism. Regarding truth and its opposite, the lie, we have to lie a little bit every day[29].

Telemedicine participants *generally* perceive the stakes of the patient's temporality and sensitivity and the concept of respect that these two modalities of being presuppose. The epistemological review of the moral economy of telemedicine therefore does not aim to show how telemedicine has moved away from the individual in the dual temporal and emotional dimensions, but rather to reassess, from an integrative ethical perspective, the place of these two dimensions in the requalification of the order of the individual in the context of telecare.

## 3.8. A renewed sense of patient experience

DEFINITION 3.6.– The *experience of the patient* with chronic disease who is monitored and assisted at a distance refers to the person's experience in their triple relationship to the disease, caregivers and care technologies. It is therefore necessary to distinguish here, on the one hand, the human and interpersonal dimension of this experience, and, on the other hand, the main effects on the person of the technological mediation that takes place within the framework of remote care and its follow-up.

29 Source: http://philosophie.initiation.cours.over-blog.com/article-l-erreur-de-kant-a-propos-de-l-imperatif- categorique-65868969.html. Page consulted on July 12, 2015.

The use of telecare implies prior recognition of the patient's experience and emotional states. As the examination of the evaluations and experience feedback collected for this book shows, taking these conditions into account makes it possible to resist the objective temptation of a technologizing medicine. A review of the evaluations on chronic disease telecare shows, in particular, how the medical institution generally characterizes the "patient experience". This is part of a *normative* and *objective* construction of the subject whose phenomenological analysis points out the limits (reductionist argument). The status accorded to the "rational body" in telemonitoring protocols and the emphasis placed on the *efficient use* of technology and *good practices* seem to overshadow the subjective dimension of the person being monitored.

As patients themselves testify, the recognition of their psycho-emotional experience remains confidential within remote care protocols. The experience – in the intimate sense of the illness – and the affects of the person play an essential role in the process of subjectivation of the pathology and the construction of a norm of self. The epistemic value of this experience, as subjective temporality, and the value of the affects of the monitored person can benefit from the insights of the phenomenology of perception and philosophical psychology. These have highlighted the relevance of the recognition of these values in the moral economy of the caregiver–patient relationship based on unconditional respect for human dignity, the sick person and their perception of illness as knowledge.

Broader consideration of these values by health professionals and caregivers, in a context of increased technicalization of care and advanced patient empowerment, has a threefold advantage: to respond to a strong need for emotional support and compassionate listening, to contribute to the development of a more efficient interhuman therapeutic alliance, and to renew the meaning of *the* patient experience (including *user experience*, see Box 3.2) in the sense of *valuable* knowledge at the heart of a telecare protocol. Examination of the multiple "time patterns" of telepatients (see Chapter 5 of this book), followed by an analysis of individual interviews with patients with chronic diseases engaged in telecare protocols (see Chapter 6 of this book), will support these initial conclusions.

The quality of *the* patient experience is intrinsically linked to the efficiency of technologies implemented in remote care and conditioned by *usability*, *user experience*, and the use of *user interfaces* (see Box 3.2).

**Usability**[30]

Usability, or suitability for use, is defined by ISO 9241-11 as "The extent to which a product can be used by specified users to achieve specified goals with effectiveness, efficiency and satisfaction in a specified context of use". It is a concept close to that of affordance, or even ergonomics, that is however broader. The usability criteria are:

– effectiveness: the product enables its users to achieve the expected result;

– efficiency: achieves the result with less effort or requires minimal time;

– satisfaction: comfort and subjective evaluation of interaction for the user.

**User experience**[31]

The user experience, often contracted by parataxis into user experience or abbreviated into UX (*User eXperience*), is a concept that appeared in the 1990s, following the publication of Donald Norman's book *The Design of Everyday Things*, to indicate that the analysis of the functions of an artifact (tool, technical object, human–machine interface) is not sufficient to understand the relationship of users with this artifact and that it is necessary to take into account the subjective and affective dimension of the person, which involves the notion of pleasure. The notion of user experience is intended to complement that of usability. The user experience is less strictly functional since it emphasizes the emotional impact cumulated with a rational benefit.

**User interface**[32]

The user interface is a type of computer interface that allows a user to operate the machine. It coordinates human–machine interaction and is part of industrial design.

**Box 3.2.** *User experience, user interface and usability definitions*

The "user experience", in the sense of the perceived interaction with an artifact (such as an internet-connected device) cannot, however, be exclusively functional: it jointly emphasizes "the cumulative emotional impact with a rational benefit" (see Box 3.2). While the effectiveness and usability of technologies are therefore critical for quality remote monitoring, they cannot overshadow the "human" dimension of patient interaction with new healthcare technologies. This interaction is not neutral: it is a psycho-emotional negotiation of the patient with objects and other remote care devices. In order to determine as closely as possible the specificities of this negotiation and its implications in terms of the redeployment of the sick subject *over time*, through uses, it is necessary first to explain the nature of human–machine interactions, in the context of remote care and monitoring, by identifying the main characteristics of the practices and processes implemented.

30 Source: https://en.wikipedia.org/wiki/Usability.

31 Source: https://en.wikipedia.org/wiki/User_experience.

32 Source: https://en.wikipedia.org/wiki/User_interface.

4

# The Patient Experience under Telemonitoring

*A pathology can be said to be objective, by reference to the doctor who practices it. But the pathologist's intention is not that their object is a matter emptied of subjectivity.*

Georges Canguilhem [CAN 10, pp. 156–157]

Three home telemonitoring pilot projects for chronic diseases have been conducted in France over the past 10 years: Diabéo (diabetes), Calydial (renal failure) and SCAD (heart failure – *Suivi cardiaque à domicile*, clinical cardiac follow-up at home). Table 4.1 provides an overview of the main benefits of these pioneering programs, which have contributed significantly to the growth of telemedicine in France and the deployment of three of its main applications: telediabetology, home dialysis and telecardiology[1].

| | **Diabéo** | **Calydial** | **SCAD** |
|---|---|---|---|
| Main sources of gains identified | Best glycemic result at 6 months<br>Hospitalizations avoided<br>Complications avoided<br>Time saving for the patient | Hospitalizations avoided<br>Saves medical time<br>Gain in patient quality of life | Improved health indicators for heart failure at 3 and 6 months<br>Rehospitalizations avoided<br>Patient comfort<br>Saves medical time |

1 These three priority areas have been chosen here because they already have champion projects with a proven level of evidence. Source: http://www.ticsante.com/Telemedecine-le-comite-strategique-de-filiere-vise-1-million-de-patients-telesuivis-en-2020-NS_2324.html.

| | | | |
|---|---|---|---|
| Population studied | 180 patients | 25 patients | 150 patients |
| Assessed medico-economic gain | 2.9 hours of transport time avoided per patient<br>2.4 hours of patient work time saved by the patient<br>Number of hospitalizations avoided during assessment | 10 fewer hospital days per year | Net savings of 538,738 euros per year |
| Participants | CERITD | Centre associatif Lyonnais de dialyse (Calydial) | CHU de Caen, URCAM Basse-Normandie |

**Table 4.1.** *Major benefits of Diabéo, Calydial and SCAD solutions*[2]

## 4.1. Telediabetology

DEFINITION 4.1.– Diabetes is caused by excessive blood sugar. Insulin normally regulates glycemia (blood sugar level). Diabetes is marked by insulin deficiency or resistance to the action of this hormone. The body becomes unable to use glucose (sugar) as an energy source. This accumulates in the blood instead of being absorbed by the cells, causing hyperglycemia (sudden increases in blood sugar levels). A blood sugar level above 1.26 g/L in the morning on an empty stomach or above 2 g/L in the daytime indicates diabetes. There are two types of diabetes.

*Type 1 diabetes*, also known as insulin-dependent diabetes, usually starts in childhood or adolescence. It is an autoimmune disease that destroys the islet cells in the pancreas that secrete insulin. *Type 2 diabetes*, which is associated with being overweight, is the most common. The pancreas still produces insulin but insufficiently and less effectively (insulin resistance). Overeating and lack of exercise can lead to type 2 diabetes. According to the WHO, France is predicted

2 *Observatoire paritaire des métiers de l'informatique, de l'ingénierie, des études et du conseil* (OPIIEC) [French joint observatory of IT, engineering, studies and consulting professions]. *Étude sur les technologies de l'information au service des nouvelles organisations de soin. Création de valeur engendrée par le secteur des TIC santé en France*, 2012, p. 116.

to have nearly one million more diabetics by 2030. Diabetes can also develop during pregnancy. This is called gestational diabetes. It requires increased monitoring (risk of hypertension in the mother and complicated childbirth) and usually disappears after delivery[3].

Telediabetology refers to the telemonitoring of patients with diabetes (types 1 and 2). One of the flagship programs in France in this field is "Diabéo" (see Box 4.1).

*Diabéo* is a remote support system for patients with type 1 and 2 insulin-treated diabetes, launched in 2004. It arose from the meeting between a research institute in diabetology, the CERITD (*Centre d'études et de recherche sur l'intensification du traitement du diabète* – French center for diabetes treatment intensification studies and research), chaired by Dr. Guillaume Charpentier, and a medical solutions and device developer, the Voluntis company. It consists of an active blood glucose monitoring logbook, accessible to patients via an application on their smartphone and a dedicated web portal, and to the healthcare team via a secure web portal. The patient records their own daily blood glucose data, as well as the amount of sugar ingested during the meal and the physical activity performed.

These data are stored and analyzed in a personalized way, and the system generates recommendations allowing them to adjust their insulin dose. The healthcare team monitors the patient's pathology remotely, in real time and accurately (telemonitoring) and can set up medical or paramedical teleconsultations with the patient. In September 2011, Voluntis and CERITD announced a tripartite collaboration agreement with Sanofi, one of the world leaders in diabetes treatment, for the development and future large-scale dissemination of Diabéo. A large-scale medical-economic study was launched in 2012 (Telesage study), including 750 patients for 2 years, in 12 regions.

**Box 4.1.** *Diabéo program description*[4]

Diabéo aims to achieve optimized glycemic control in patients by intensifying treatment to prevent complications, incidents and accidents. The device concerns patients with type 1 diabetes. Between 2000 and 2009, the prevalence rate of diabetes in France continued to rise. It has even progressed faster than expected. In 2009, it reached 4.4% according to the *Institut de veille sanitaire* (INVS) [French institute for public health surveillance], which means 2.9 million people affected. Type 1 diabetes accounts for 5–10% of diabetes cases observed.

3 Source: http://www.topsante.com/medecine/maladies-chroniques/diabete#tab2. Page consulted on November 23, 2014.
4 Extract from the Telemedicine 2020 report [TEL 12, p. 10].

| Key figures for diabetes (types 1 and 2) in France and worldwide |
|---|
| (census 2016) |
| Europe: 60 million |
| France: 2.9 million |
| United Kingdom: 3.6 million |
| United States: 29 million |
| Global annual mortality rate: 3.4 million |

**Box 4.2.** *Key figures for diabetes in France and worldwide*[5]

This form of the disease can occur at any age, but most often it occurs in childhood or early adulthood. In France, it is commonly accepted that nearly 200,000 people have type 1 diabetes. The average age of type 1 diabetic patients in France is about 41 years. The implementation of treatment is cumbersome and burdensome for the patient based on measurements and adjustments four to five times a day. Treatment is lifelong. The patient must learn:

– to carry out their capillary glycemia measurements several times a day to evaluate the quantity of carbohydrates in their food;

– to adjust their doses of rapid insulin before each meal, according to these two elements;

– to correct a possible blood sugar divergence;

– to adjust the insulin dose according to the physical activity being planned for;

– to adjust the basic insulin dose (slow insulin or base pump flow rate) active at night and between meals.

The CERITD, a non-profit-making association created on July 6, 2006, aims precisely to improve the prevention and treatment of diabetes. It is based at the CHSF (*Centre hospitalier sud francilien*), in Dr. Charpentier's endocrino-diabetology department. It contributes to the public service's support for improving diabetes treatment via the development of insulin pump treatment, and the active participation of CERITD medical and paramedical teams in the management of inpatients and outpatients. CERITD helps to disseminate knowledge about diabetes

---

5 Source: World Health Organization, http://www.euro.who.int/en/health-topics/noncommunicable-diseases/diabetes/data-and-statistics.

to both patients and physicians through training programs. CERITD conducts research programs in telemedicine, but also runs a program involving patch, pump and artificial pancreas devices, academic programs to better understand certain aspects of the disease and research programs for the early prevention of diabetes in families with type 2 diabetes. According to CERITD, insulin-related accidents currently account for about 100 deaths per year[6].

> The research programs conducted by CERITD concern the insulin-treated diabetic patient, with the precondition that insulin treatment, which involves a lifetime of constraints that affects about 10% of all diabetics, can occur at any age, but most often appears during childhood or early adulthood. Individuals with insulin deficiency produce very little or no insulin. The goal of telemedicine in insulin-treated diabetes is to help the patient improve the quality of their treatment through better control of their blood sugar and through a precise adjustment of insulin doses (often in the context of complex insulin regimens). The objectives of telemedicine use are as follows:
>
> – improve patient compliance (defaulting by periods);
>
> – help the patient make the right decision (at the time of the injections);
>
> – optimize medical time and improve access to specialized care;
>
> – allow traceability of care;
>
> – reduce diabetes management costs.

**Box 4.3.** *Research objectives of the CERITD*[7]

The goals of telediabetology are to reduce the occurrence of diabetes complications, and to ensure patient and caregiver satisfaction. The clinical return to Diabéo validated by the Telediab 1 study showed an improvement in HbA1c of 0.9 points in 6 months (consequent impact on the occurrence of diabetes complications). HbA1c, an abbreviation for glycated or glycosylated hemoglobin (i.e. the index used to judge blood glucose control during the 2–3 months preceding a blood test), is a biological value used in diabetes monitoring. Conducted by Dr. Alfred Penfornis, in 2009, on a cohort of 180 patients with type 1 diabetes over a 6-month period, the Telediab 1 study demonstrated the economic (savings on transport costs, reduction in the number of trips) and organizational (gain in medical productivity, better control without an increase in medical time) benefits [PEN 12].

6 Source: www.ceritd.fr.
7 Source: www.ceritd.fr.

Satisfaction indicators are regularly collected from patients and caregivers involved in the protocol using an anonymous satisfaction questionnaire sent to all patients (results published in 2012). Such a survey should be conducted annually to analyze the quality of the cooperation protocol and improve patient care. In addition, indicators have been put in place since 2011 to assess the effectiveness, usefulness and cost of implementing the protocol.

Management of insulin-treated diabetic patients, with electronic glycemic log book type solutions (used in the context of Diabéo) allows:

– for the patient: real-time assistance in adapting insulin doses, a strengthened link with the healthcare team, personalized follow-up, improvement of glycemic data (HbA1c);

– for healthcare teams: better time management, implementation of an educational approach, assistance in optimizing patient treatment, contributing to better glycemic control and prevention of complications of the disease;

– for the scientific community: creation of patient databases;

– for health insurers (primarily health insurance): a tool to strengthen the application of good practice recommendations, an additional means to better prevent complications of diabetes, particularly cost elements, especially in terms of hospitalizations;

– for the healthcare industry: assistance in the proper use of treatments (antidiabetic drugs and blood glucose meters for glycemic control assistance).

**Box 4.4.** *Benefits of Diabéo for stakeholders*[8]

The implementation of telecare in the context of Diabéo is mainly based on a conventional visit, a face-to-face initiation visit and telemonitoring. During a conventional visit, the delegating diabetologist proposes to maintain their face-to-face consultations and to add a follow-up of blood sugar levels, their HbA1c, hypoglycemic incidents, hyperglycemic potential by telemedicine. In other words:

– by offering to load an electronic blood glucose diary into their personal smartphone to help them decide how to calculate their insulin doses;

– by explaining to them that their glycemic data, unit portion, physical activity and insulin doses will be transmitted via a secure website to themself or, if they accept it, within the framework of a protocol of cooperation to a nurse chosen by the patient, trained by them, and to whom they will delegate the monitoring and the remote analysis of their

8 Source: www.ceritd.fr.

glycemic balance and, in the event of glycemic imbalance, this delegate will call them by telephone to prescribe an adjustment of their insulin doses;

– by giving them a consent form.

The patient then creates a patient profile (marital status, type of treatment, glycemic goals: fasting, postprandial, at bedtime). They send the information to the delegate who gives them an appointment for the so-called face-to-face initiation visit. During this visit, the delegate:

– has them sign the consent form;

– opens an initiation visit in the electronic shared patient file that allows the traceability of all delegating actions visible to all 24 hours a day, 7 days a week and 365 days a year;

– appropriates by validating the profile data at two levels (their diabetes and their treatment);

– trains the patient to use the electronic log book (diagnosis on physical and psychological adherence and training in the functional insulin therapy (FI) program loaded into the software). The visit takes place according to a therapeutic education approach submitted to the ARS Ile-de-France, centered on active listening, which is done in three stages:

- educational diagnosis with the patient,

- assessment of the patient's own acceptance of the disease and/or material,

- setting realistic and achievable improvement goals by the patient themselves and the choice of insulin doses.

**Box 4.5.** *Monitoring protocol with Diabéo*[9]

Finally, Diabéo is equipped with a telemonitoring platform enabling telemedicine nurses, acting on behalf of the general practitioner, to interact in real time, "at the right time" with the patient, according to the evolution of results. Then, if the bad results persist, Diabéo puts the patient in contact with their doctor for pharmacological reinforcement, and possibly the latter, in contact with a diabetologist of the network, in order to limit "therapeutic inertia". Because of this system, the Telediab 2 evaluation, conducted following Telediab 1, mentioned above, showed a significant difference of at least 0.36% on HbA1c at 12 months and a significant decrease of 19% in hospitalizations in 1 year. The improvement in the quality of life of patients with diabetes, thanks to Diabéo, is further confirmed by

9 Source: www.ceritd.fr.

recent evaluations of other telediabetology systems such as, for example, Educ@Dom.

The recent evaluation of the Educ@Dom program, carried out by the *Unité de soutien méthodologique à la recherche du CHU de Toulouse* (Toulouse university hospital center research methodological support unit), IRIT laboratories, ENSEEIHT MEDES (*Institut de médecine spatiale*), the DIAMIP Care Network and SADIR, leads to conclusions substantially similar to those of the evaluations from Telediab 1 and 2:

> *Our study aims to evaluate the telemonitoring management of blood glucose control and hygiene in T2D patients, compared to standard management. The main objective is to compare the effectiveness at 1 year of these two treatments via HbA1c. The secondary objectives are the maintenance of efficiency at 2 years, the evolution of knowledge, hygiene and dietary behavior, quality of life, clinical and biological parameters [...] The home monitoring device includes therapeutic education software on a touch pad and connected sensors: blood glucose meter, balance and actimeter. Data from the sensors are sent to a secure data center and are synthesized into a telemonitoring report available online for physicians who decide what action to take. Results obtained: the first qualitative results on test patients demonstrated the feasibility of the device. The use of the device for 3 months motivated them to improve their lifestyle. The nutrition software helps them to balance their diet and wearing the actimeter encourages them to walk more* [SCH 14].

DEFINITION 4.2.– Postprandial blood glucose (PPG) is the blood sugar level usually checked 1 h 30 min to 2 h after the start of the meal.

Following the Telediab 1 study, two additional programs were implemented in the Diabéo system: one allowing the automatic analysis of glycemic data to allow caregivers to better motivate their patients and/or adjust their treatment accordingly, based on the results transmitted; the other program is the integration of a computerized delegation of tasks protocol established by the physician according to the individual needs of the patient and offering the physician specialized nursing time to assist them in the training and follow-up of their patients. In practical terms, patients have an active electronic log book that can be downloaded onto a tablet, lent to the patient, where blood sugar levels are recorded. The patient is encouraged by automatic personalized messages to act from one day to the other on their bad PPG, or at the end of a half-day, by reducing their prandial carbohydrate intake or by increasing their physical activity. If they receive basal insulin, they will automatically be offered a titration according to their fasting blood glucose (FBG) results. A secure website has also been created where patient log book data are automatically transferred and accessible at any time to authorized caregivers, then analyzed, generating automated alert messages, in the event of persistent blood glucose misuse or insufficient or incorrect system usage.

| Domain | Application | Objective | Reference |
|---|---|---|---|
| Type 2 diabetes | WellDoc Diabeo-2 | Compliance – demographics | 2 |
| Type 1 diabetes | Diabéo MeOS | Decision support – reduce follow-up – compliance | 3 |
| Gestational diabetes | DiabGest | Facilitating/procedure monitoring | x |
| Diabetic retinopathy | OphDiab | Mass screening demographics | y |
| Diabetic foot pathologies | AirPeDia | Regional organization of specialized care | In progress |
| New technologies | CareLink DiabeLoop | Ensuring treatment compliance | Design |

**Table 4.2.** *The six fields of telediabetology in France [BEN 13b, p. 25]*

In order to request reimbursement of the solution, a larger scale medico-economic study (12 CHU coordinators in the regions, 200 hospital and liberal diabetologists and six telediabetology nurses in the regions) were convened following discussions with the *Haute autorité de santé* [French high authority for health]. Covering a cohort of 750 patients over a 2-year period, the Telesage study was launched in France in 2012 with Sanofi-Aventis as sponsor. This study includes patients with type 1 and type 2 diabetes who are chronically out of balance despite basal-bolus insulin treatment. It should make it possible to precisely evaluate the importance of the metabolic benefit obtained, according to whether Diabéo is managed alone or in cooperation with specialized nurses within the framework of a protocol of delegation of tasks. The costs of this monitoring modality will be assessed, as will the achieved savings. The results of this study should enable the health authorities to decide on the financing of the Diabéo system, the efficient organization of care around the Diabéo system and its dissemination on the national territory.

## 4.2. Autonomous home dialysis telemonitoring[10]

DEFINITION 4.3.– Dialysis is a supplemental treatment that provides only incomplete renal function replacement. It removes excess waste and water (or toxins) accumulated in the body from the blood. It is associated with constraints, in particular those to submit to dialysis sessions and to a strict diet, limited in water, salt, potassium and phosphorus intake.

Renal failure affects more and more people in France. Because there are few symptoms, it is most often diagnosed at a terminal stage that requires immediate dialysis: a surrogate technique that uses a machine to filter blood instead of failing kidneys. Dialysis is usually done in a specialized clinic with an average of three weekly sessions, each lasting 3–5 h. A very restrictive treatment for the patients and expensive for the French national health insurance system. According to a recent report by the *Cour des comptes* (Court of auditors), the costs of dialysis are as follows: € 53,028 per year for non-assisted peritoneal dialysis (by a nurse); € 87,036 for hemodialysis in a medical center. To compare with those of the transplant, which amount to € 53,273 the year of the kidney transplant, then only € 13,536 the following years [COU 15].

**Key figures for France**

– Ten percent of adults have kidney dysfunction. The number of people affected is estimated at nearly 3 million;

– by the end of 2014, nearly 80,000 people were being treated for end-stage renal disease (ESRD);

– 56% were on dialysis and 44% had a kidney transplant;

– in 2014, nearly 11,000 new patients were treated for ESRD with a 2% increase over the previous year. One in two patients was 70 or older;

– in 2014, 3,241 kidney transplants were performed.

**Box 4.6.** *Key figures for renal failure in France*[11]

10 In France, there were an estimated nearly 42,000 kidney dialysis patients in 2016. But only 8% of them benefited from home dialysis. Health insurance calls for the development of this technique, which is less restrictive for the patient and more economical for the healthcare system. Source: http://www.allodocteurs.fr/maladies/prostate-reins-vessie/insuffisance-renale-dialyse/la-dialyse-a-domicile-une-technique-a-developper_14749.html. Page consulted on November 12, 2015.

11 Source: http://www.ameli-sante.fr/maladie-renale-chronique/quest-ce-que-la-maladie-renale-chronique. html. Page consulted on September 10, 2015.

**Key figures for Europe**

– 3.7 million patients treated for end-stage renal disease;
– 6% growth per year;
– 3 million patients are treated with hemodialysis or peritoneal dialysis;
– 750,000 kidney transplants were performed (2016 census).

**Other countries**

(Number of patients per million inhabitants – 2016 census)

Taiwan: 3,940
Japan: 2,720
United States: 2,180

**Box 4.7.** *Key figures for end-stage renal disease in Europe and worldwide [FRE 16]*

Owing to the high costs of dialysis, new treatment methods were introduced in clinical practices. Teledialysis, for example, consists of the implementation of communicating systems between a "referral" center where the team of nephrologists is located and a satellite unit where the patients and the paramedical team are located. This modality for the management of patients suffering from chronic renal failure was created in the late 1990s and has been the subject of experimentation in various countries, notably in Canada and France (Centre Hospitalier de Saint-Brieuc). In France, the publication in January 2010 of a report by the *Haute autorité de santé* on "Conditions for the implementation of telemedicine in medical dialysis units" [HAU 10] has provided a framework for teledialysis deployment. These recommendations deal in particular with organizational, legal, ethical and geographical aspects.

From a technical point of view, the teledialysis system in a medical dialysis unit, as it is offered to patients today, is composed of three components which, when used simultaneously, allow:

– telemonitoring by viewing and storing dialysis session parameters from the generators;

– audiovisual communication between the team of nephrologists, the nursing team and the patients of the remotely monitored unit;

– use of support applications necessary for remote medical procedures: computerized patient records, teleprescription and possibly an expert data analysis system to assist the nephrologist in their decision-making.

**Box 4.8.** *Teledialysis device in medical dialysis units in France*[12]

12 https://fr.wikipedia.org/wiki/Télédialyse. Page consulted on September 10, 2015.

Over the years, other practices have been developed in France and other countries of the world, such as *telemonitored autonomous dialysis at home*. More comfortable for the patient, this can be accomplished in two ways. The first is based on the hemodialysis technique. Similar to dialysis clinics, *hemodialysis* consists of filtering the blood outside the patient's body, using a machine installed at home. The second is *peritoneal dialysis*, which involves injecting a solution into the body of the patient with chronic end-stage renal disease (ESRD), which is then evacuated. In both cases, a medical and paramedical team remains permanently available and ensures regular follow-up of the patient and equipment. These techniques remain very poorly developed in France: only 8% of dialysis patients benefit from them because the organization of care is based on specialized clinics (see Box 4.9 for the situation in other countries).

**Worldwide key figures for hemodialysis at home worldwide**

(as a percentage of total number of dialysis patients treated)

New Zealand: 18%

Australia: 9%

Canada: 3–6%

Western Europe: 3–6%

United Kingdom: 15%

France: 8%

**Box 4.9.** *Worldwide key figures for hemodialysis at home [WAL 17]*

The French national health insurance calls for the development of home dialysis to promote patients' autonomy, but also to save money[13]. The duration and frequency of daily dialysis vary according to the degree of renal failure of the patient: it can be 2 h a day or 30 min several times a day or 10 h in a row for night dialysis. One of the pilot telemonitoring programs for dialysis patients at home in France is Calydial (see Box 4.10).

13 The cost of a self-dialysis session is 200–250 euros while the cost of a hemodialysis session in a clinic is 340–380 euros. Source: http://www.allodocteurs.fr/maladies/prostate-reins-vessie/insuffisance-renale-dialyse/la-dialyse-a-domicile-une-technique-a-developper_14749.html. Page consulted on October 10, 2015.

Calydial is a telemonitoring system for patients with end-stage renal disease undergoing home peritoneal dialysis, in existence since 2005. Wishing to limit the loss of medical time linked to a very high number of home visits, the Lyon dialysis community center (Calydial) approached Orange Healthcare, which had the communicating pen technology (data reception, sending and storage). At each dialysis (three to four times a day), the patient ticked off on a preformatted sheet the information concerning the progress of their dialysis session (type and weight of dialysis tubing).

This information, as well as data on their state of health (weight, blood pressure and medication), was transmitted to the healthcare team once a day due to the communicating pen technology. The latter recorded the data entered on the sheet using an integrated microcamera and sent them to a server, which centralized all information from all patients and stored them on a website dedicated to Calydial's healthcare professionals. This procedure was gradually replaced by a health application used on a tablet (see ap'Telecare description in section 5.3.1 of this book). The data are analyzed by an expert system that generates a therapeutic decision aid based on an alert system.

**Box 4.10.** *Calydial program description*[14]

Calydial (*Centre Associatif Lyonnais de Dialyse*) is a 1901 law association created in 1984. It is a private health establishment of collective interest (ESPIC) adhering to the FEHAP[15], specialized in the treatment of patients with renal failure, management of chronic ESRD and chronic non-dialyzed renal disease (ND-CKD). Calydial provides two types of activity: dialysis, on the one hand, and primary and secondary renal failure prevention, on the other hand, at two kidney health clinics. Patients at high cardiovascular risk are referred by general practitioners, cardiologists and diabetologists in private practices and hospitals. Its region of activity covers the south of Lyon and Vienna[16]. The Calydial home peritoneal dialysis telemonitoring project was based, until 2013, on a validated medical protocol carried out according to the steps listed in Box 4.11.

14 Excerpt from the Telemedicine 2020 report [TEL 12, p. 11].

15 ESPICs (*Établissements de santé privé d'intérêt collectif*) [private health establishments of collective interest] supplant the institutions participating in the public hospital service (PSPH), at the end of the vote of the Hospital, patients, health and territory act (HPST). ESPIC covers the private non-profit sector. The majority of these institutions depend on the *Fédération des établissements hospitaliers et d'aide à la personne* (FEHAP) [Federation of hospitals and personal care institutions]. They are financed in the same way as public hospitals and managed by a legal person under private law. Source: http://www.hospimedia.fr/actualite/mots_cles/espic. Page consulted on November 2, 2015.

16 Five projects have been carried out over the last 10 years by Calydial: (1) telemonitoring at home (version No. 1): communicating pen; (2) the setting up of a dialysis unit monitored remotely in a hospital center (Lyon South Hospital Center); (3) the telemonitoring at home on a PC; (4) since 2013, telemonitoring at home (version No. 3); and (5) the ap'Telecare project.

– At home, the patient writes down simple daily clinical information (weight, blood pressure, blood sugar levels) concerning their state of health and the progress of their dialysis on standard, rasterized sheets every day[17].

– At the same time as writing, a *communicating pen* with a microcamera records and stores this information.

– This information is digitized and transmitted (once a day) by Bluetooth to a mobile phone and arrives in real time, via the GSM network, to a secure remote server. It collects and centralizes all information from all patients at home, which is accessible in real time at Calydial's call center, from its intranet and under conditions of use. This call center is a professional on-call service 24 hours a day, 7 days a week.

– Because of a software adapted and developed within Calydial, all this information is analyzed and generates alerts on medical protocols, thus making it possible to detect early clinical changes (weight gain, hypertension) and to anticipate complications (water overload, hypertension, emergency problems, etc.). A registered nurse (RN) monitors up to 25 patients at a time, at least once a day (about 15 minutes a day).

– The medical monitoring protocol, developed within Calydial, is based on a system of graduated alerts; it makes it possible to decide what action to take according to risk situations and dysfunctions in the patient and/or paramedical practices analyzed. Either everything is fine, or there is a decision to be taken (corrective actions to be taken before clinical decompensation), and this can either be taken by the RN or someone with medical competence must intervene, contacted by telephone by the RN.

– The exchanges are traced in the alert management application and in the computerized medical file, via the report of the action.

– A consultation takes place every 2 months at the patient's home.

**Box 4.11.** *Telemonitoring protocol for home peritoneal dialysis [AGE 12, p. 98]*

Since 2013, Calydial has implemented a software application that replaces the communicating pen while remaining faithful, in broad outlines, to the medical protocol described in Box 4.11. Called ap'Telecare, an innovative software solution on touch tablet, it is adaptable to any home monitoring project. It is a secure application from an authorized health data host. Like the communicating pen, ap'Telecare allows the patient to be taken care of from their home and allows them to become involved in the autonomous management of their protocol. Indeed, ap'Telecare makes it possible to connect, with an interactive mode, the person being monitored and their healthcare team. The constants collected are communicated in real time to the doctor who can check the alert thresholds and modify the established

17 The focus of this chapter is limited to the evaluation of the medical project and the organizational aspects of telemonitoring of renal failure at home. For a detailed discussion of the effects of home dialysis on the telemonitored person, see Chapter 6 of this book.

protocols at any time. The home dialysis patient can record their daily data on the touch screen of their tablet and communicate with their doctor by message or teleconsultation. Ap'Telecare includes the patient file on a dashboard, personalized protocol management (for example frequency and duration of dialysis sessions), videoconferencing and messaging capabilities, and alarm threshold management. The ap'Telecare solution, therefore, considerably improves the patient's quality of life compared to the communicating pen used until 2013. However, no study on temporal (constraint) or emotional factors (positive and negative emotions) in the management of ap'Telecare or any other "companion software" for patient telemonitoring at home has been published to date. The qualitative survey of patients with chronic renal failure for this book, the results of which are discussed in Chapter 6, is specifically designed to assess this dimension.

The term "companion software" refers to medical software applications installed on smartphones or tablet computers that track side effects, guide symptom management, promote adherence to treatment protocols, recommend doses and promote data sharing with medical teams. In addition to the ap'Telecare software, developed by TMM Group[18], there are today many companion softwares for medical telemonitoring developed by other French companies or start-ups (ICARE, H2AD or BeeTree).

The evaluated benefits of Calydial's home dialysis telemonitoring solution and ap'Telecare are listed in Box 4.12.

– Very good patient acceptability;

– gain in organization;

– improvement of the quality of care via intervention before complication;

– better monitoring of education with identification of risk situations;

– limitation of home travel for health professionals;

– new organization based on "validated" medical and paramedical surveillance protocols;

– routine integrated practice in the operation of the establishment and in the patient's home telemonitoring.

**Box 4.12.** *Main benefits of Calydial for home dialysis telemonitoring [AGE 12, pp. 103–104]*[19]

18 www.tmm-groupe.com.

19 These results are confirmed by the testimonials of chronic patients interviewed for this book (see Chapter 6 of this book).

## 4.3. Telecardiology

DEFINITION 4.4.– A medical device designed to be implanted in whole or in part in the human body or placed in a natural orifice is called an implantable medical device (IMD). The definition of a medical device is given for the European Union, in Directive 93/42/EC, amended by Directive 2007/47/EC. The transposition into the French public health code in 1994 gives the definition in article L5211-1[20].

Telecardiology refers to the telemonitoring of patients with an IMD as part of the management of a cardiac pathology (treatment or diagnosis): pacemaker, implantable cardiac defibrillator or long-term rhythmic holter. Sixty thousand pacemakers and 8,000 defibrillators are implanted in France for approximately 65 million inhabitants. The cardiologist has three types of electronic prostheses for the diagnosis and treatment of heart rhythm disorders: pacemakers, defibrillators and implantable holters that are called IMDs. The severity of the pathologies is not the same according to the prostheses implanted:

– 80% of patients implanted with a pacemaker do not have an associated heart disease;

– 85% of patients implanted with a defibrillator have an associated heart disease of which approximately 15% have a severe rhythm disorder;

– these prostheses can be the object of a punctual telemonitoring with calendar and/or continuous event-dated deadlines.

Telecardiology systems operate differently from each other (daily versus regular transmission, transmission by GSM network versus wired mode and exclusive transmission of prosthesis data versus complement by blood pressure or weight data). Depending on the manufacturer, the devices offered are more or less automated (transparent use for the patient). Training adapted to each system is offered by the academic research communities [AGE 12]. Several studies have already assessed the benefits of telemedicine in monitoring IMDs (pacemakers and cardiac defibrillators) and the benefits for the patient of telecardiology (see Box 4.13).

– Improved treatment follow-up and quality of care;

– fewer trips with time savings and improved quality of life (fewer inappropriate shocks);

– earlier diagnosis of problems and prevention of complications;

20 Source: https://fr.wikipedia.org/wiki/Dispositif_médical.

– a reduction in the length of hospital stays and the use of care (hospitalization, consultations);

– a decrease in the number of face-to-face consultations but access to specialized medical advice; a decrease in transport for the patient;

– reduced isolation for implanted patients who are far from an implant center: elderly patients, difficult and expensive transport.

**Box 4.13.** *Main benefits of telecardiology [AGE 12, p. 73]*

The COMPAS study between 2005–2009, for which the CHU Haut-Lévèque de Bordeaux was an investigation center, was funded by Biotronik. This is an 18-month, prospective, randomized study on a pacemaker with a comparison of two branches of 150 patients each (branch 1: classic follow-up; branch 2: telemedicine follow-up). The primary objective was to analyze the impacts on the number of cardiovascular deaths and hospitalizations. The other objectives were to evaluate the effects on the number of consultations and the medico-economic aspects; 2010: American TRUST study on defibrillators (study of the delay between the beginning of the rhythm disorder and the medical intervention); 2011: American CONNECT study: study on defibrillators with the objective of analyzing the impact of telemonitoring on the number of consultations and on the duration of hospitalizations; 2011: the publication of the first results of the multicenter study ECOST (Pr. Kacet, Lille) on patients with DMI Biotronik demonstrated the reduction in the number of inappropriate shocks delivered to ICD patients; 2011: publication of the results of the EVATEL study (in which the CHU Haut-Lévèque is an investigative center), multicenter and multiconstructors (four different manufacturers). Randomized study on 1,501 patients, implanted in 31 different French clinics, with a follow-up over 12 months. The objective was to compare telemedicine monitoring with conventional monitoring, particularly on the occurrence of major cardiovascular events.

**Box 4.14.** *Clinical studies conducted in telecardiology in France and the United States*

Chronic heart failure is a major public health problem: first cause of hospitalization after 60 years (220,000 hospital stays per year in France), 120,000 new cases per year, significant mortality (31% survival to 5 years) and very high cost (12–20,000 euros per patient per year) [ESC 14]. Compared to other conditions, such as cancer or myocardial infarction, heart failure is a disease of particular concern (as shown in Figure 4.1). Heart failure in France represents 210,000 hospitalizations per year for 60,000 patients. The number of people suffering from heart failure is increasing, due in particular to the ageing of the population, as the disease is strongly linked to age. Rehospitalization rates are very high. An estimated 40,000 cardiac patients are monitored remotely at home today[21].

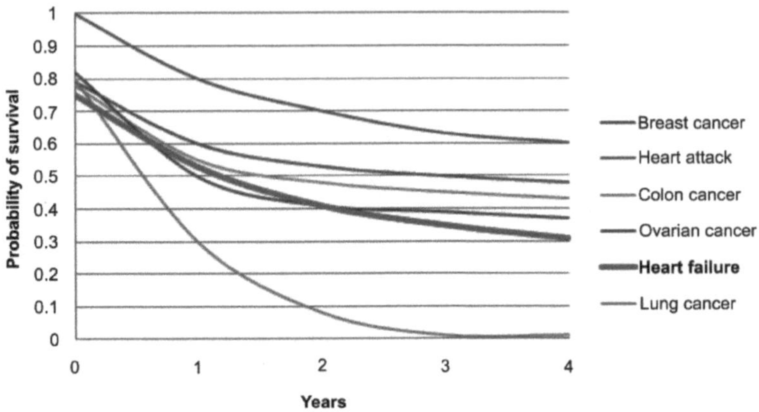


**Figure 4.1.** *Probabilities of survival by disease type[22]. For a color version of this figure, see www.iste.co.uk/bardy/telemedicine.zip*

| **Heart failure in France and worldwide** |
|---|
| (census 2016) |
| France: 1 million patients[23]<br>*40,000 patients fitted and remotely monitored* |
| Europe: 15 million patients [KEL 15, p. 3] |
| United Kingdom: 7 million[24] |
| United States: 5.7 million[25] |
| One million patients diagnosed each year worldwide |

**Box 4.15.** *Key figures for chronic heart failure in France and worldwide*

21 Source: http://docplayer.fr/7949240-Table-ronde-1-telesurveillance-de-patients-atteints-d-insuffisance-cardiaque.html. Page consulted on November 12, 2015. See also, on this point, the presentation by Professor Régis Beuscart, biostatistics hospital practitioner, head of the information sector and medical archives department, Université Lille 2, CHRU of Lille, source: http://www.catelparis.fr/remuneration.html. Page consulted on April 1, 2016.

22 Source: www.cardiauvergne.com. Page consulted on January 1, 2015.

23 Source: https://www.fedecardio.org/Les-maladies-cardio-vasculaires/Les-pathologies-cardio-vasculaires/linsuffisance-cardiaque.

24 Source: British Heart Foundation. https://www.bhf.org.uk/research/heart-statistics.

25 Source: Centers for disease control and prevention. Heart Failure Fact Sheet. https://www.cdc.gov/dhdsp/data_statistics/fact_sheets/fs_heart_failure.htm.

One of the pioneering telecardiology programs deployed in France is SCAD, of which Box 4.16 provides an overview.

> SCAD is a project to improve the management of heart failure deployed in Basse-Normandie in 2006, which notably includes the implementation of a home monitoring system for patients discharged from hospital. It is the result of a joint reflection between the cardiology department of the CHU de Caen and the URCAM Basse-Normandie. The system allows an exchange between the patient and a remote care team. It is based on the use of a dedicated touch-screen terminal (web-e-phone) installed in the patient's home.
>
> The patient enters daily health monitoring information (weight, blood pressure, edema, dyspnea [shortness of breath], cough, etc.), and consults advice and information on hygiene and diet rules and compliance with treatment on the terminal screen. It also allows direct communication between the patient, the healthcare team and the attending physician, because of a classic telephone functionality and an e-mail functionality. The healthcare team accesses all the data entered by the patient via a secure internet portal, which generates alerts in the event of an emergency. Preliminary results from a study conducted in 2011 suggest a 40% reduction in hospital days after 3 months of SCAD follow-up. However, the cost of a single day of hospitalization in cardiology is higher than that of the device (€ 900 per patient).

**Box 4.16.** *SCAD[26] program description*

Initially, SCAD makes it possible to carry out an educational follow-up to avoid hospitalizations by reinforcing the patient's therapeutic education. Therapeutic education begins at the hospital with the establishment of an educational diagnosis setting negotiated goals between patients and caregivers (disease imperatives, *desiderata* and patient acceptance). This is followed by individual interviews and workshops on the pathology, treatment, physical activity, dietetics and warning signs for early consultation. Then a telemedicine device of SCAD is installed in the patient's home for a duration of 3 months [AGE 12, pp. 88–89]. The system is offered to all patients in a condition to receive therapeutic education and they are warned that it is not an emergency device.

This device is based on the remote collection and transmission of data by means of a dedicated touch-screen terminal allowing the entry and reading of medical information. All the information exchanged is analyzed by a nurse manager at the hospital center to which the patient belongs. Based on the indications given by the patients, messages are sent to them, generated automatically by a follow-up algorithm. The latter was developed by a medical team from the University Hospital

26 Extract from the report Telemedecine 2020 [TEL 12, p. 11].

of Caen (Dr. Annette Belin and Dr. Rémi Sabatier) and paramedical Basse-Normande in collaboration with the Canadian team of NewIT Santé, which developed the device. The terminal also allows direct communication by telephone or e-mail between the nurse, the patient and the attending physician. SCAD provides a daily patient follow-up and education relationship based on a validated clinical algorithm that includes:

– health monitoring parameters (weight, edema, dyspnea, cough, blood pressure);

– appropriate advice and information on hygiene and dietary rules and compliance with treatment;

– advice and incentives for further home exercise retraining.

The system is not restrictive for the patient, who accesses it whenever they like during the day. The patient is questioned by the terminal on a weekly cycle[27]. The system provides three alarm levels with an intervention protocol (see Box 4.17).

– Yellow alarm: call the patient, verification of the veracity of the recorded results, consultation advice with the doctor and verification the next day that the patient has called. If this has not been done, put them in touch with the doctor;

– orange alarm: call the patient, check the veracity of the results entered and call the doctor so that they can see their patient quickly;

– red alarm: call the patient, assessment of degree of urgency and call to patient's physician. If they cannot be reached or cannot respond with urgency, call 15 (emergency number in France).

**Box 4.17.** *Alarm system for SCAD [AGE 12, p. 88]*

– Voluntary approach as an extension of therapeutic education workshops (patient agreement essential, information letter and consent);

– identification of patients with chronic heart failure (CHF) at high risk of rehospitalization;

– monitoring the critical post hospitalization period;

– hospital-city link around the patient;

– service delivery around the patient;

– monitoring data useful for treatment optimization;

27 Vital signs: Monday, Wednesday, Friday; dietary hygiene rules: Tuesday, Thursday; exercise training: Saturday; and psychological follow-up: Friday evening.

– contact between the hospital team and the doctor;

– reassuring monitoring after hospitalization;

– learning through repetition ("training");

– acquisition of daily self-monitoring;

– improving adherence to boards and standardizing practices;

– evaluation of practices (patient satisfaction questionnaires, quality of life, knowledge of the disease, etc.);

– early or preventive consultation triggered;

– involvement of the liberal practitioner in SCAD follow-up.

**Box 4.18.** *Evaluated benefits of SCAD [AGE 12, pp. 88–89]*

The other expected benefits of SCAD and the major telecardiology programs currently deployed in France (such as OSICAT, PIMPS (*platforme interactive médecin patients santé*, Interactive physician patient healthcare platform), or Cardiauvergne) are a significant reduction in readmissions (30–40% minimum), deaths (20–30%) and costs to society (30% per patient per year). The OSICAT study is a large multicentric study conducted under the auspices of the Toulouse University Hospital Center and the Midi-Pyrénées ARS, and in partnership with APETCARDIOMIP (*Association pour l'éducation thérapeutique en Midi-Pyrénées*, Association for the therapeutic education in the Midi-Pyrénées) and the various URPS (*Union régional des professionnels de santé*, French regional union of health professionals). The objective is to compare morbidity and mortality among heart failure patients after 18 months of telemonitoring with a telecardiology program. A total of 870 patients will be included, comparing two groups (telemonitoring versus standard management) of patients with CHF. The management mode studied consists of an alternative outpatient telemonitoring strategy in which patients will benefit from telephone follow-up by a specialized nurse and the monitoring of weight and symptoms by telemonitoring [GAL 14].

The objective of the PIMPS program coordinated by Dr. Jourdain of the Therapeutic Heart Failure Unit at René Dubos Regional Hospital Center in Pontoise is to demonstrate the impact of a telemonitoring solution on patients with heart failure, including professional training and educational reinforcement, telemonitoring and the therapeutic follow-up of patients[28]. The Cardiauvergne service is defined as follows by Roman Eschallier:

28 Source: www.numerique-sante.fr.

> *Cardiauvergne, a telemonitoring and care coordination service, has set for itself the challenge of improving prognosis while keeping patients out of hospital, thus ensuring a better quality of life. This health coordination group (GCS, 35 partners) began including patients in January 2012: today, 745 of them benefit from this solution financed by the Auvergne ARS and local authorities, and developed thanks to a partnership with Almerys (Orange subsidiary specialized in medical IT data management). Two principles have contributed to the system: health professionals remain "masters of the game" and telemonitoring must be simple for it to be acceptable to both patients and professionals. Thus, the patient is equipped with only one sensor: an Internet-connected scale. Independent nurses with smartphones provide information on clinical progress (functional signs, edema, heart rate, blood pressure). Biological monitoring is automatically included in the record by the analytical laboratories, and the dispensing pharmacist informs the therapy each time the prescription is dispensed. The computerized patient file can be consulted by their professional who identifies themself with their HPC card [health professional card]. All data is analyzed by an expert system that generates alerts or alarms according to the severity level set by terminals. These are managed by the coordination cell within 3 hours, 7 days a week (1.5 cardiologists, 2 nurses qualified in therapeutic education, and 1 medical secretary). Alerts and alarms initiate a call to the patient, who will receive therapeutic education advice, and if the patient's condition justifies it, a call to the attending physician, or even the cardiologist for therapeutic adaptation. An evaluation was done at the end of 2 years of operation (355 days of average follow-up) on the first 558 patients included. The death rate is 12% per year (against 28% to 35% depending on the series with conventional management), while the readmission rate for new heart failure attacks is reduced to 13.6% per year (against 26% to 40% in European cohorts). Finally, the economic impact is remarkable (savings estimated at € 7,000 per patient per year). Patient and professional satisfaction is unanimously favorable* [ESC 14].

## 4.4. Medical telemonitoring: an objectifying practice?

Several lessons can be learned from the review of patient experience evaluations in the context of telemonitoring at home. On the one hand, behind these evaluations there is a normative construction of the patient experience, in a relationship to oneself dominated by technique: that of an *objectified experience*. On the other hand, this standardized construction is to be seen as *conditions of experience and relationship to others*, technomedically determined. In other words, it is essentially a perception of limited telemedical experience. The assessment of the patient experience under the term "quality of life" is generally based on the use of standardized questionnaires. But by integrating the subjective dimension of the patient as a quantifiable metric (standardized questionnaires can indeed evaluate the stress level, the intensity of depressive episodes, the recurrence of anxiety)[29], telecare systems perpetuate the status of the rational body, i.e. the body as the "object" of (medical) reason that *calculates* and *projects* according to precise objectives of well-being and self-improvement. The philosopher Isabelle Queval proposes an enlightening analysis of the rational body for this analysis: "If the rational body is calculated, drawn and programmed, it is *a priori* according to relatively identifiable ends, starting from no less identifiable means. Formulated in this way, the hierarchy of means and ends refers to a first definition of rationality generally inspired by the *optimism of the Enlightenment*, i.e. *modern*: reason oriented toward success, or reason guided by progress and helped in this by technology" [QUE 08, p. 82]. When fitted, or even "instrumented", the sick body is numbered, quantified and subjected to clinical telemedical examination for objectification purposes. The normative construction of the individual by telemonitoring techniques can be perceived, under these conditions, as medical reductionism: the objective telemonitoring of the disease as a biological "mechanism" is preferred over the knowledge derived from the patient's subjective perception, which constitutes their experience. In short, home medical telemonitoring seems to value the measurement of the *objectifiable state* of the disease as an end in itself. According to this mechanistic vision, the body must be repaired. The body must be increased in its biomechanical potential. However, this objectivist reductionism has several effects. On the one hand, it contributes to biasing evaluations of the patient experience by obscuring the subjective dimension of the person being monitored and the value of the "patient knowledge" resulting from this experience to the benefit of objective self-knowledge, as we have seen. On the other hand, by focusing on the objectifiable state of the sick body, telemonitoring protocols and their evaluations can miss certain "black spots" in the patient experience, in particular:

– the time-consuming and restrictive use of communicating telemonitoring objects;

– the impact of telemonitoring on the daily pace of life;

– stress and anxiety factors related to the technical *and* human aspects of telemonitoring (caregiver–patient relationship, support or lack of emotional support);

– individual feelings about the disease, its course and risks;

29 A questionnaire of this type is presented in Chapter 3 of this book.

– the perception of present and future time through the use of communicating devices;

– the effects of the psychological experience of the illness and the telemonitoring system on the caregiver–patient relationship.

Home telemonitoring, as it can be perceived in evaluations and feedback, thus seems to be in line with the trend toward the objectification of the "sciences of pathology" of the 20th Century, from which Georges Canguilhem situates the methodological and ethical challenge of modern medicine, which refers, by analogy, to the objectivist ambition of scientific research described by Bernadette Bensaude-Vincent:

> *The ideal of objectivity that has animated scientific research for centuries is really to eliminate all biases, to avoid all contamination by "human factors" – this is what the American philosopher Thomas Nagel and the science historian Lorraine Daston describe as "the view from nowhere"* [BER 09, p. 116].

The objectification specific to the sciences of pathology called into question by the possibility of subjectivation also refers intricately to the opposition between a Kuhnian conception of science, aiming at objectivity and reproducibility, and a Poperrian vision, admitting refutation. For Georges Canguilhem, author of the *Écrits sur la médecine* [CAN 02] [Medical writings], it is important to affirm and clarify the humanist and relational vocation of medicine against its objectivist reductionist temptations:

> *It is always the relationship to the sick individual, through the clinic, that justifies the pathological qualification. While acknowledging the importance of objective methods of observation and analysis in pathology, it does not seem possible to speak, in any logical correction, of "objective pathology" [...] The [objective] intention of the pathologist is not that their subject is a matter emptied of subjectivity* [CAN 10, pp. 156–157] .

Despite telemedical practices of objectifying the patient and advances in technologies and objective methods of telemonitoring and analysis, the telepatient cannot be *reduced* to purely material objectivity. In line with the ethical requirement of recognition of the person as a subject, claimed here by Canguilhem, it is essential, according to the philosopher Philippe Barrier, to take into account the individual as a whole: "The disease affects a *subject*, and not a simple organized living matter, a somato-psychic being, an irreducible chiasm to one of its components, a being somehow bi-polar whose *soma* and *psyche* would be the extremities" [BAR 10a,

p. 160]. This remark draws attention to the fact that the patient is above all a subject worthy of respect in their entirety and in their human integrity.

The attention given to the mediation of the disease by connected devices interposed, in the context of telemonitoring, should not be to the detriment of the sick person (telemonitored patient). The sick person is not a simple "neutral" and docile living matter that remains immobile in the face of illness or care. In reality, the (tele)patient, as a subject, is essentially driven by "self-care". Less objective than functional, the management of this "concern for oneself" by medical telemonitoring is however problematic.

The emphasis placed in telemonitoring on the rational benefits of the patient's experience, in the sense of a "functional biological system", argues in favor of the use of technologies intended to enhance the ability of the telemonitored patient to *care for themself* from a normative point of view, i.e. to comply with the *technical normativity* of the care protocol (for example self-managed use, under medical telemonitoring, of the cycler or the insulin pump). Nevertheless, the relative subjugation of the person to technical normativity, as a reductionist mode of care, must not make us forget the human capacity, biologically human, to face the disease, as Barrier points out here:

> *With the addition of the reflexive pronoun ("to heal oneself"), it is also to oneself that the care will be directed. This particular attention to well-being and suffering will be able to be brought, in a way, to oneself as well as to another, thus adding a last dimension to care: to heal oneself, for example medically speaking, is also to care for oneself from a normative point of view, it is to ensure that the corrective processes of treatment operate as well as possible and to favor natural self-regulatory processes* [BAR 10a, pp. 155–156].

Patient compliance with care technologies – now also referred to as *compliance* (see definition on p. 24) – seems to have the effect of "functionalizing" self-care and, incidentally, "objectivizing" the person within a protocol. However, it cannot mask the "subjective" specificities of the care experience, which raises a major question: what status should be given to non-objective aspects of the disease such as the patient's feelings and affects? Health or illness assessments, in the context of telecare, seem to name an objectifiable state of the body, not a relationship of the person to an environment or an intimate relationship to the illness (temporal and emotional).

# 5

# The Person Standing the Test of Digital Clocks

*It is to be prepared for the unexpected, not to plan for everything.*

Juliette Favez

QUESTION 5.1.– How does telecare manage patient temporality and sensitivity? What concepts of temporality and sensitivity does it construct?

Reporting on the telemonitored patient's temporal specificities implies, on the one hand, shedding light on the objects and practices used today to monitor patients with chronic diseases at home and, on the other hand, identifying the patient's temporal patterns shaped by these telecare practices.

In this perspective, we propose to identify here the specificities of "high points", i.e. the key moments of a home telemonitoring protocol for the patient: time of care and follow-up with caregivers/helpers, time of interaction with the computerized device, the communication device, time of data entry, transmission of biological data, etc. This insight into the *temporal distribution* that takes place within the framework of a telecare protocol, centered on chronic disease, makes it possible jointly to draw up a typology of telemonitored patients and, in the last instance, to outline the broad lines of the telecare temporal paradigm, which is currently divided into two distinct modalities.

On the one hand, the telecare system allows caregivers and patients to communicate with each other, or with internet-connected devices, at regular intervals, which has the main effect of building and "harmonizing" a temporality of follow-up: all participants in the telecare relationship can indeed have access to the same medical information (in real time, depending on the pathology). But these real-time telemonitoring practices often require the patient's active participation, which can be experienced as a constraint, and therefore weigh on their day-to-day management and their "intimate" temporality, even leading to certain "phase shifts". Studies show in particular that the patient's experience, determined by the illness, and the caregiver's experience mainly determined by the care, do not necessarily coincide in the context of home telemonitoring. These two contradictory modes of time experience (harmonization and phase shifts) are not, however, the product of chance or the sole result of practices imposed by the telecare system. Harmonizations and phase shifts are also, from a phenomenological point of view, the result of practices that arise from multiple body perceptions and multiple perspectives on the illness. It is therefore necessary to bring out, beyond the telepatient's "temporal landscapes"[1], the values carried by the computerized management of the person's temporality and illness and, conversely, to identify the values of the patient linked to their corporeal experience within the framework of the telecare relationship.

## 5.1. Synchronized participants in a network

A home medical telemonitoring system allows three types of participants to be connected: a patient, a telehelper (nurse, medical advisor, assistant) and a telepractitioner (see Figure 5.1).

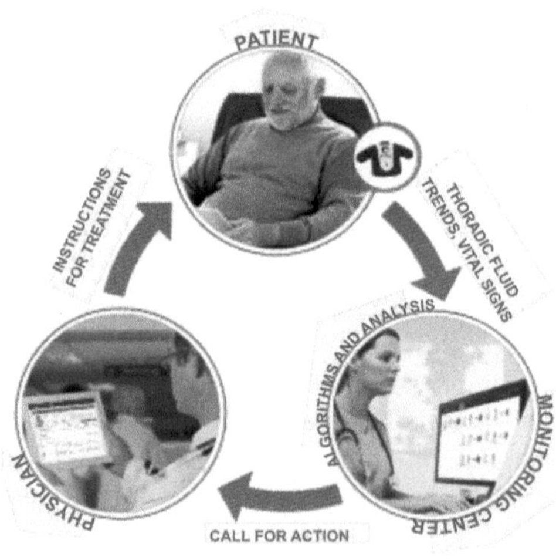


**Figure 5.1.** *Typical organization of a home medical telemonitoring system*[2]

1 This expression is inspired by the following work: [ADA 98].
2 Source: http://healthtechinsider.com/tag/monitoring/. Page consulted on July 8, 2015.

This synchronized and networked connection of participants as well as permanent access to telemonitoring data is made possible because of the various communication devices and computerized systems used by patients, telepractitioners and/or caregivers (smartphones, tablets, personal computers, Wi-Fi terminals, cameras, IT servers). In the case of the cardiac patient, data are transmitted continuously to the telepractitioner via an implanted defibrillator that communicates with a relay terminal installed in the person's home. Patients with kidney failure or type 1 or 2 diabetes may also use certain so-called *smart* devices (internet-connected biosensor, thermometer or actimeter) in addition to their home medical equipment (dialysis device or insulin pump).

These new digital health technologies allow patients to be telemonitored (without intervention on their part for heart patients) to communicate with the various participants in the telecare relationship to enter and transmit data on their state of health to receive electronically medical advice, and/or a recommended dosage, and/or a prescription and to carry out themselves, if necessary, the care recommended by a practitioner. It should be noted that caregivers and patients share confidential medical data, because of these digital technologies, on a dedicated and (in principle) secure telemonitoring platform (see Figure 5.2).

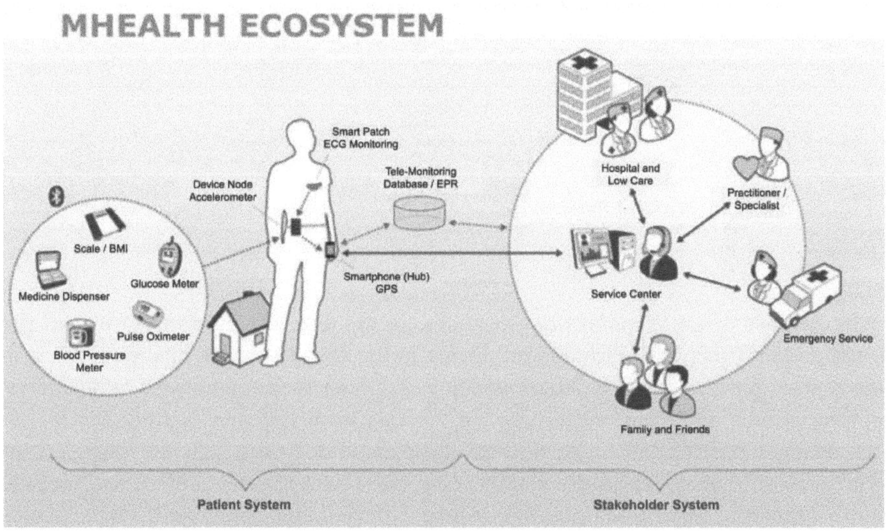


**Figure 5.2.** *Computerized model of a home telemonitoring system*[3]

3 Source: http://www.mdtmag.com/article/2013/05/wireless-enabled-remote-patientmonitoring-solutions. Page consulted on July 8, 2015.

In the context of telemonitoring of renal failure, cardiac failure and diabetes, telepatients are assisted in adapting their treatments (for example their insulin doses in the case of diabetes) according to their needs or assisted in managing certain difficulties related to their treatments. These are based on objectives defined by healthcare professionals and use real-life data reported by the patient using digital technologies. Home telemonitoring thus makes it possible to optimize individualized care for the sick person. This management is done through the transmission of medical data that is translated into practicle terms by regular feedback to the patient (medical advice, dosages, recommendations), made before and after the analysis of their data by the telemedical team. These different stages of patient telemonitoring take place in two main ways: synchronous or asynchronous (see Figure 5.3).

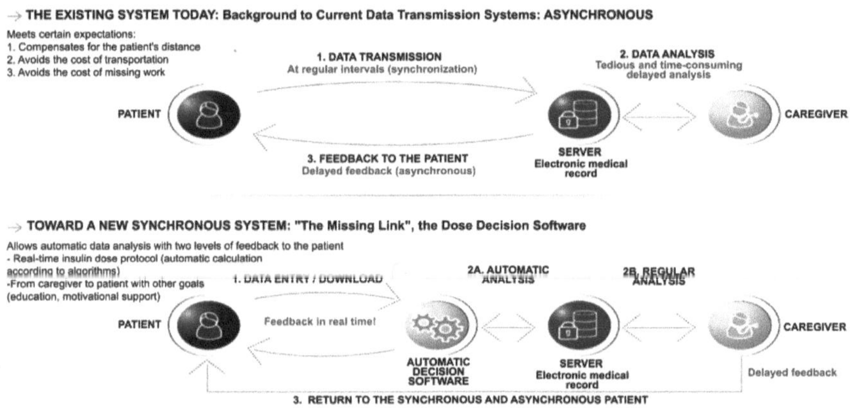


**Figure 5.3.** *The existing system today and future systems: from asynchronous to synchronous [LEU 14]*

In the case of an *asynchronous* connection, the data are transmitted in real time by the patient (or by their implanted device in the case of the cardiac patient) to the data center. Feedback to the patient is delayed. Under these conditions, data analysis is often tedious and time consuming. In the case of a *synchronous* connection, the data are first transmitted to an automatic decision software that performs a "first level" real-time feedback to the patient; then these data are transmitted and recorded in the patient's electronic medical record before being processed. Synchronous connections are now becoming more widespread because they respond more quickly to the needs of patients and telecare providers; the latter have constantly updated data that enable them to operate improved "second level" feedback, in the context of teleconsultation in particular.

Whatever the mode of connection (synchronous or asynchronous), the benefits of telemonitoring are numerous for the patient: closer ties with the doctor; reduction in costs, and less frequent travel to undergo medical examinations; reduction in costs related to missing work; increased empowerment; closer networking with those involved in the telecare relationship (caregivers, family and social care providers). Whatever the chronic illness, the daily tasks to be performed by the patient, which are essential for follow-up, involve not only implementing but also accepting a shared routine (or even a ritual) with the caregiver, the terms of which are negotiated in advance with the project instigators.

In summary, the synchronous and asynchronous connections between the participants in the telecare relationship are articulated around three "high points" that punctuate the patient's daily life:

1) *Manual entry and transmission of biological data by the patient (or automated transmission in the case of the implanted cardiac patient).*

2) *Delayed feedback (asynchronous connection) or real-time feedback (synchronous connection) to the patient BEFORE data analysis by the telemedical team.*

The practitioner's feedback to the patient varies according to the patient's state of health. In addition, automatic decision software can replace the telepractitioner, at this stage, by transmitting certain programmed recommendations or, in the case of the diabetic patient, by calculating the recommended insulin dose according to the data transmitted by the patient.

3) *Telepractitioner's feedback to the patient, delayed, AFTER data analysis.*

Automated alerts can also be sent to the patient or practitioner at this stage. These alerts inform stakeholders about the evolution of the illness by enabling them to anticipate health risks. Key elements of the telecare preventive strategy, data entry, data transmission, real-time or delayed feedback, and alerts to the patient and the practitioner are part of a disease-management process whose *multiple temporalities*, which seem not to have been the subject of any qualitative study to date, are listed, as an indication, in Box 5.1.

| **Telecare objects and systems** | **Individual** | **Practitioners and helpers** |
|---|---|---|
| *Real-time assurance*<br>Intercommunications between telecare participants<br>* | Disease experience (physical and psychological) under telemonitoring<br>* | *Caregivers*<br>Time of the one-on-one conference with the telepatient<br>* |

| | | |
|---|---|---|
| Entry and transmission of patient biological data to the healthcare team<br>*<br>Automated analysis of physiological data<br><br>*Provide synchronous connections*<br>Feedback from the medical team to the patient (or automatic decision software in the case of diabetes)<br>*<br>Real-time alerts<br><br>*Provide asynchronous connections*<br>Feedback and/or delayed alerts (by email on tablet or SMS on smartphone) to the patient (after data analysis)<br>*<br>*Ensure a continuous follow-up of the person* | Time of recording and transmission of body data (temperature, actimetry, etc.)<br>*<br>Time of care (at home or in hospital)<br>*<br>Compliance time<br>*<br>Time for dialog with a health professional, other patients (in the case of so-called "expert" patients) and/or a helper (social or family)<br>*<br>Time of the one-on-one conference with a (tele)practitioner<br>*<br>Time for wellness activities associated with telecare protocol (yoga, tai chi, etc.) | Time of the transmitted data management (reading and interpretation)<br>*<br>Time for diagnosis and/or personal advice<br>*<br>Time of communication with other professionals and/or third-party helpers<br>*<br>Patient communication time<br>*<br>Time to develop the telemonitoring protocol<br><br>*Helpers*<br>Time for dialogue with the patient<br>*<br>Time of accompaniment (material, psychological) |

**Box 5.1.** *Telecare time regimes*[4]

Home telemonitoring has the advantage of keeping the person in their environment, thus avoiding trips to the hospital or to the doctor while guaranteeing quality medical monitoring. In addition, through therapeutic education, the patient has the possibility of taking up, under strict medical supervision, all or part of the care at home: blood glucose reading and insulin dose injection after analysis of their data for the diabetic patient, peritoneal dialysis for the patient with renal failure and medication for the cardiac patient. This patient empowerment represents a considerable technological advance and a real therapeutic gain for the person whose quality of life is greatly improved. However, the delegation of care to the patient

4 Only the temporality of the person being monitored is examined in this chapter.

raises numerous ethical problems, including the risk of preferring autonomy to the detriment of the human dimension of the care relationship[5]. Involving the patient's adherence to a follow-up protocol that engages their responsibility, in close consultation with the telemedical team, the delegation of care also constrains the patient to a discipline focused on the management of daily activities that are quite time consuming: body data recording (temperature, actimetry – i.e. measuring physical activity – blood sugar levels), data recording and transmission, care, taking medication, dialog with a telepractitioner or a helper and wellness activities possibly associated with the telecare protocol (relaxation, meditation, massage).

To manage medical telemonitoring and its *chrono-diversity* (i.e. the plural temporality regime of the participants involved), a precise organization is therefore established in the medical centers, and the roles are distributed between nurses, patients and practitioners. As shown in Figure 5.1, the standard organizational model for home telemonitoring of chronic patients is based on "simple" telemonitoring of chronic disease, with recording, transmission, data analysis and alert generation. However, this model tends to evolve rapidly today, particularly in the treatment of renal failure and diabetes. For patients with these chronic diseases, the progressive implementation of a connection with their specialist physician, using a telemedicine-mentoring interface in their hospital – and in the future at home – represents a concrete example of this technological evolution [BER 15b]. Moreover, while this typical telemonitoring model is based on a dynamic and "high points" that punctuate the daily life of the telemonitored patient, it nonetheless points to a determining element of the operational logic of telemedical monitoring, namely the control of the temporal factors of chronic disease. It is indeed essential, from the (tele)medical point of view, to take these factors into account to define the patient's profile, in real time, and to determine as closely as possible the evolution of their state of health, in accordance with the preventive logic of telecare.

Thus, the diabetic patient's telemonitoring is concretely translated by the recording of blood sugar levels before and/or after meals. These simple readings or "snapshots", collected at regular intervals by the patient, are a digital representation of their illness in real time. This reconfiguration of the long-term condition (LTC), in the form of digital data that are sensitive to body temporality, constitutes a key element of the telecare protocol, or more precisely a "chronoscopy" of the telemonitoring; in other words, a "time image" of the body that materializes the "digital exteriorization of the perspective" on the disease and its temporality. Far from being insignificant, this exteriorization of the medical perspective, theorized in

5 The main ethical issues involved in empowering the person being monitored are evaluated in the third part of this work (see Chapter 8, "'Chrono-Sensitivity': From Concept to Ethics").

its time by Michel Foucault [FOU 04], is transformed, in its digital variant, into a perspective that is sensitive to biological rhythms and that concretized by the digitization of the disease and its temporality, "seen from within". In other words, the externalization of the perspective on the disease in the context of telemonitoring finds its foundation in the "numbering" of the body's intimacy that takes place through connected devices that stimulate in the telecare relationship, according to the type of disease concerned, a differentiated rhythm: in "real time", "shared time", or "synchronized" mode. Examination of these multiple temporalities makes it possible, on the one hand, to define a typical telepatient profile for each chronic disease by shedding light on the main characteristics of the "telemedical perspective" of the disease and its temporality and, on the other hand, to measure the epistemological scope of the values and issues related to telecare that are sensitive to the temporality of the person. To support this dual analysis, we propose to examine here, beforehand, the temporal specificities of the telemonitoring of the pathologies submitted to the study and to describe the operational modalities of the technologies that make this type of monitoring possible today.

## 5.2. The cardiac patient: "real-time" data producer

Heart failure is a condition characterized by heart slowdowns called bradycardia or critical accelerations. When the heart beats too slowly, the patient suffers from syncopations. When it beats too fast it endangers the patient's life with the risk of sudden death. Heart failure affects elderly people in particular.

Several types of devices can now be used to monitor the heart patient at home: pacemakers (for bradycardia), implantable defibrillators (to prevent sudden death) and devices called "resynchronizers", more complex devices that improve heart failure symptoms. In addition, implantable monitors *without therapeutic action*, unlike pacemakers and defibrillators, can be used for the diagnosis of severe syncope whose suspected etiology is of cardiac origin. All these computerized and communication devices are, in most cases, implanted under the clavicle and connected to one, two, or three probes, depending on the complexity of the prostheses, and placed inside the heart cavities[6].

Telemonitoring of these cardiac prostheses in combination with conventional cardiological monitoring by the local cardiologist has two advantages. It allows the patient to be monitored on a daily basis via their prosthesis thanks to programmable periodicity transmissions and transmissions linked to "alerts". On the other hand, it

6 For a detailed presentation of cardiac defibrillators, see Chapter 6, "Experiential Knowledge of the 'Subject of Care'".

reduces the frequency of medical consultations. A single annual face-to-face consultation can even, in certain cases, be maintained in order to clinically monitor the patient, the implant box and make any adjustments to the prosthesis.

In most hospitals with a telecardiology platform[7], a trained nurse gives the patient the transmission box after the implantation of their cardiac device[8]. The patient receives instructions to connect this box to an electrical outlet, preferably on the bedside table, as well as explanations of the (quite simple) operation, and carries out the first transmission. The nurse registers the patient on the telemonitoring website and then enters the pathology and the treatment. In agreement with the doctor, the nurse programs the transmission frequency (for example monthly) and adapts the alerts according to the patient's pathology. During the night, the implanted device will communicate with the telemonitoring box and send the data to a secure site via a GSM system. As for all medical sites, the CNIL (*Commission Nationale de l'Informatique et des Libertés*) (French National Commission on Informatics and Liberties) has verified the high level of security of this site, which is accessible by login and password.

The data received are divided into technical data specific to the implanted prosthesis and clinical data such as the occurrence of tachycardia. These data are automatically prioritized into different severity levels. The next day, the doctor in charge of the telemonitoring receives in their email all the alerts, identified according to their severity level. The data are checked, and if necessary patients are called for management. In some hospitals, the management of transmission mails and site surveillance are done by a doctor. The doctors delegate to the paramedical team the email reception with a first sorting carried out according to the degree of urgency, and then the paramedical team turns to the doctor for the continuation of the management.

From a technomedical point of view, the cardiac patient with an implanted communication device (i.e. not involving any manual intervention of the patients, who can however use other communication devices to report biological data such as weight, actimetry and temperature) is, with the transmission box installed at home[9],

---

7 We are inspired here by the experiment conducted at the Henri Mondor Hospital in Asnières. http://chu-mondor.aphp.fr. Page consulted March 1, 2015.

8 The focus here is exclusively on the latest generation of implantable communication devices, such as those designed by Sorin as part of the Smartview solution proposed by Orange Healthcare, which automatically record the patient's cardiac data and then transmit them directly to the data center located in the hospital environment.

9 Some defibrillators, such as the Saint-Jude models, do not require the use of a cardiac data transmission box because the cardiologist records the data on the integrated memory card for later reading.

the single "interface" of the telemonitoring network. From the network point of view, in fact, the cardiac patient is a *measuring device*, whose indications must be properly filtered so that only information concerning certain aspects of health is conveyed to the telemonitoring team (arrhythmias, blood pressure). The patient is a *producer of real-time, mobile data*, which must be kept in contact with a transmission network at all costs: in the context of "informed consent", the patient gives delegation powers to the network administrators. On the other hand, the patient, as a producer of real-time data, is a source of information which, if correctly correlated and interpreted, must enable a decision to be made on their state of health. It is also a database entry, to which potentially considerable amounts of useful or irrelevant information are associated in the case under consideration; but it is an entry that must be able to be reliably linked to the natural person if necessary. The patient is therefore a geographical point that must be reached by all means in case of emergency.

This functional approach to the monitored cardiac patient raises many legitimate questions about the inhumanity of a medicine of technological progress. On this particular point, patients, doctors and informed telemedicine commentators seem to agree. Thus, Jacques Marceau, President of Aromates and co-founder of the *Collectif santé numérique* (Digital Healthcare Collective), wonders if telemedicine is not in danger of becoming "a medicine in which the doctor will have become an executor at the orders of protocols and statistics", or if telepatients are not soon to become "data producers in the service of a system governed by big data"[10]. This argument refers to the criticism of evidence-based medicine (EBM) mentioned earlier (see Chapter 2). Despite these concerns based on the fear of dehumanizing and dehumanized medicine, which must be taken into account in the reexamination of the values of humanity and solidarity inscribed, through the Hippocratic oath, at the heart of (tele)medical practice, it nevertheless seems important to briefly recall the clinical benefits of telecardiology from a temporal point of view.

10 "It may seem paradoxical to ask ourselves the question of the place of the human in a discipline that touches the human in the depths of itself: medicine. And yet, this question is being asked of us today, both by technological progress and by the increasing constraints, particularly economic constraints, on our health system and the organization of care. Are we moving towards dehumanized medicine? Process medicine? A medicine in which the doctor will have become an executor at the orders of protocols and statistics? Where the patient will be constantly controlled, forced, 'coped' by all kinds of connected tools and sensors that will creep into their own flesh? Will we become connected objects? Data producers at the service of a system governed by big data?" Excerpt from the opening speech by Jacques Marceau, president of Aromates and co-founder of the *Collectif Santé Numérique*, at the conference *Numérique en santé : quelle place pour l'humain?* organized on Wednesday, March 12, 2014 as part of the 5th *Assises des Technologies Numériques de Santé*. Summary of the conference, p. 17.

*First, the cardiac incident can be "anticipated", in some cases, with the implantable monitor used in home monitoring.*

A patient who goes into atrial fibrillation (a rush at the level of the auricles), clearly identified by their prosthesis but not felt by the patient, risks a blood clot forming in the heart, which if it migrates into the brain will be responsible for a hemiplegia type stroke. To prevent this risk, the patient should be promptly called for a consultation or hospitalization to introduce anticoagulants to prevent the formation of this clot. Telemonitoring will make it possible to be reactive, detect the clot and treat the patient early. Telemonitoring is not an emergency system since there is an incompressible delay between transmission and data management, but it is a means of improving the speed of patient management, sometimes upstream of symptoms, thanks to the alert system that can be implemented through communication devices. Telesurveillance of medical devices implanted in association with certain major industrial players in the treatment of heart failure, such as Sorin or Medtronic Assistance for example, as well as medical teams and patient associations, are major players in improving the quality of life of patients suffering from heart failure.

*Second, telehomecare makes it possible to refine diagnosis and to space out planned hospital visits.*

This close telemonitoring is particularly effective for patients at high risk of arrhythmia and sudden death, such as those who have already survived cardiac arrest, myocardial infarction or cardiomyopathy.

*Third, the production and analysis of biological data, because they are performed in real time, optimize the conditions of follow-up and patient care.*

If real-time management of the patient's cardiac data makes it possible to guarantee their well-being, it also contributes here to redefining the concept of time which, for Bergson, can be understood in two ways: either by consciousness or by technique [BER 12, BER 13]. The subjective time of consciousness is linked to our representations (thoughts, feelings) while objective time, that of the clock, acts as a common, universal measure of time. In the telecare context, these two temporalities tend to blur. The patient's subjective time depends on the telemonitoring system that shapes not only their representation, but also their perception and awareness of the disease. The accuracy of the data collected on their state of health and the reliability and efficiency of the telemonitoring system give the patient an unprecedented acuity on their disease. Human intuition does not seem to give way to technical truth, but it is inspired by it, because technique has proved its worth. Some recent studies show that almost all arrhythmias (i.e. disturbances of the normal heart rhythm) are now detectable in real time because of telemonitoring (see also [SOR 13]). Two conclusions can be drawn from this observation. On the one hand, the awareness of the disease is penetrated by technical temporality. On the other hand, real time invests the management of a disease fundamentally induced by a rhythmic insufficiency of the cardiac patient. Under these conditions, continuous data production through telemonitoring may appear to be a therapeutic asset. The quantity of data "produced" in real time, from the patient's point of view, or "collected", from the practitioner's point of view, seems to depend on the responsiveness of a computerized system – and therefore the quality of care – to a "corporeal system", that of the patient, which is deemed to be defective. The value of real time thus seems to be essentially constituted in a reactive therapeutic practice that confers on the patient the status of data producers: a practice that is also based on the principle of "circumvention" of the person.

In summary, physical and technical temporalities go hand in hand in the context of telecare. This juxtaposition obeys an imperative to manage the temporality of the disease "in real time" without, however, in the case of the cardiac patient equipped with a non-therapeutic implantable monitor, giving the person a grip on this personal management. The telemonitoring protocol, as we have seen previously, does not include the cardiac patient in the operational loop, either in terms of data capture or transmission or analysis. Thus, the cardiac patient is mainly a producer of real-time data (which they do not necessarily understand) whose management is ensured "by proxy" by the telemedical team. The "telemedical perspective" is, in this case, overhanging. However, there are other ways of managing chronic disease that involve the patient in a different temporal relationship to telemonitoring.

## 5.3. Renal failure or the "chrono-responsible" patient

DEFINITION 5.1.– The chrono-responsibility of the patient designates the responsibility that takes place through the technical objects to manage the key moments of the care scheme.

Renal failure is the impairment of the functioning of both kidneys, which no longer filter blood properly. The disease is called acute if the dysfunction is transient and reversible and chronic if the destruction is irreversible, with no possibility of cure. In this case, the disease can be stabilized.

Acute renal failure most often occurs after an attack such as a sudden and transient drop in blood pressure, during hemorrhage, general infection (sepsis), drug intoxication or when the urinary tract is blocked by a prostatic calculus or adenoma. The kidneys take a few days to spontaneously return to normal functioning after treatment. During this period, dialysis should be used to allow the patient to survive during the renal self-repair process. Chronic renal failure does not, by definition, decline. It is induced by pathologies (diabetes, hypertension) that progressively and irreversibly destroy the various renal structures. There are five stages of the disease up to the terminal stage at which filtration capacity is less than 15% of the normal capacity for both kidneys. This stage requires consideration of renal replacement techniques: dialysis and transplantation[11].

Unlike hemodialysis, which uses extracorporeal circulation equipment, blood purification by peritoneal dialysis takes place inside the body, within the peritoneal cavity. If there is severe kidney failure, the body gradually takes on substances that need to be removed by dialysis. Dialysis consists of bringing the blood into contact with a sterile liquid (dialysate) whose composition is close to that of plasma (the liquid that makes up about 60% of the blood) through a membrane that serves as a filter. In hemodialysis, this process takes place outside the body and the membrane is artificial. On the other hand, in peritoneal dialysis, the exchanges take place in the abdomen and the membrane is the peritoneum. In both methods, the physical phenomena involved are the same[12]. Hemodialysis is a heavy and restrictive treatment but is essential for the survival of some people. About 50% of people treated by hemodialysis have pain, related to not only kidney disease and associated diseases (such as diabetes), but also dialysis[13]. At present, research is exclusively medical on the subject.

To perform dialysis, an artificial fluid, dialysate, is introduced into the peritoneal cavity. This liquid will then be removed after a certain contact time. In the cavity, the dialysate is in direct contact with the two layers of the peritoneum. As the exchange takes place because of the permeability of the membrane, the dialysate will capture the elements present in the blood plasma that are to be eliminated (such as urea or creatinine) as well as the excess water (see Figure 5.4).

Once the liquid has been infused (dialysate), the stasis phase begins. During this phase, exchanges take place between the blood plasma and the dialysate within the cavity for a defined period of time. At the end of this period, the liquid containing the waste is drained by the catheter. The infusion/stasis/drainage cycle can be repeated several times in 24 hours depending on the medical indications and methods used.

11 Source: INSERM. http://www.inserm.fr/thematiques/physiopathologie-metabolisme-nutrition/dossiers-dinformation/insufisance-renale. Page consulted on December 1, 2014.

12 Source: https://fr.wikipedia.org/wiki/Hémodialyse. Page consulted on December 1, 2014.

13 http://www.fnair.asso.fr/remisebourse2016.htm#210316. Page consulted on June 1, 2016.

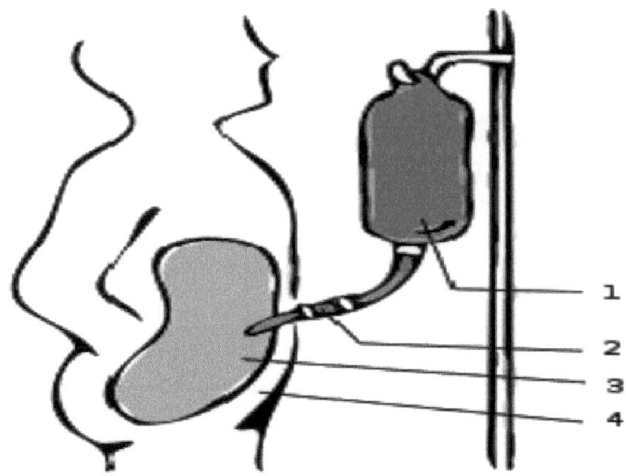


Legend: **1**: dialysate, **2**: catheter, **3**: peritoneal cavity, **4**: abdominal cavity.

**Figure 5.4.** *Schematic diagram of peritoneal dialysis*[14]

Dialysis can take place in the person's hospital or home. The manipulations (which can be carried out by the person themselves after a therapeutic education process) must be carried out under hygienic conditions which are very strictly indicated by the care protocol, in prevention of the risk of local infection or peritoneum infection (peritonitis in particular). If the person is not autonomous, they may be assisted by someone close to them or call on a home nurse. Peritoneal dialysis can be performed in two ways: continuous ambulatory peritoneal dialysis (CAPD) and automated peritoneal dialysis (APD). In France, the CAPD method is more widespread than the APD method, representing 70% of the population of patients treated by peritoneal dialysis.

### 5.3.1. *Application for prevention and monitoring*[15]

Ap'Telecare is an application for post-ambulatory follow-up and telemedical home monitoring of chronic diseases, including kidney failure. It is an application that aims at a double action: preventive (before dialysis) and follow-up (after dialysis). In both cases, it is an innovative solution for individualized management of each patient, in their home. Ap'Telecare allows you to create interactive links for

14 https://fr.wikipedia.org/wiki/Dialyse_péritonéale. Page consulted on December 1, 2014.

15 The first experiments on the follow-up of dialysis care, by minitel, were conducted in France in the 1980s at the Bordeaux hospital, by Bruno Salgues, head of mission appointed by Raymond Barre, Prime Minister.

a secure medical follow-up dialog between doctors, the healthcare team and their patients. Its ergonomics is adapted for an easy and autonomous use of the topics: information of the constants in a customized medical protocol that is arranged, as in the case of the telemonitoring of the cardiac patient, around several daily "high points" (see Figures 5.5 and 5.6).

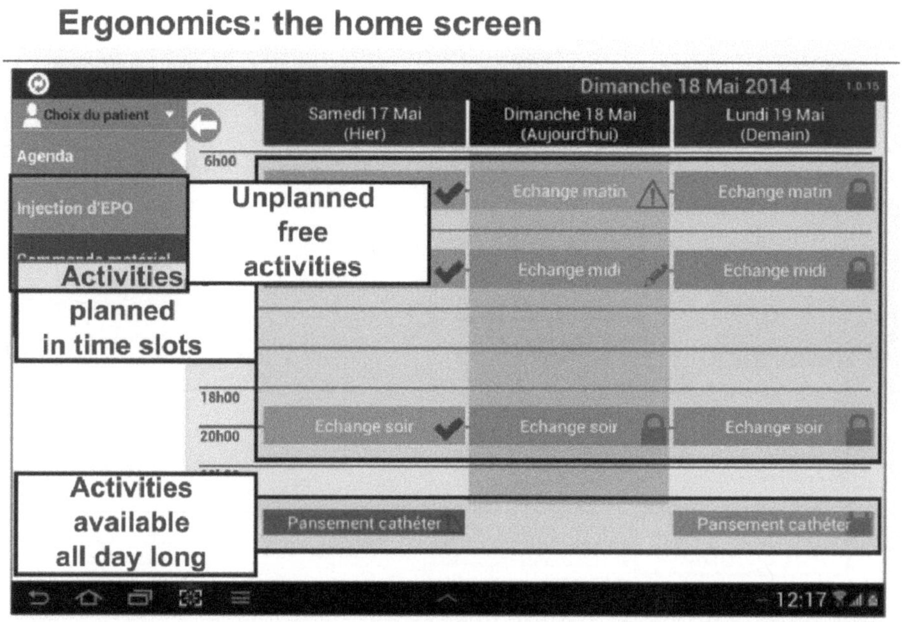


**Figure 5.5.** *Ap'Telecare application home screen (Source: [DAB 14])*

### 5.3.2. *Preventive action (non-dialyzed patient)*

A randomized pilot study was launched in December 2010 by two dialysis centers: Calydial (Lyon) and Agduc (Grenoble). This study was carried out on 80 outpatients from Grenoble and Lyon suffering from chronic renal failure without dialysis in order to verify whether telemonitoring gives better results than conventional monitoring in terms of treatment compliance, patient quality of life and reduction in adverse events. The study was conducted over six months with two cohorts of patients, one of whom received training by the medical and nursing team on the parameters to be monitored according to the pathology: patients learn to read their prescription, and to make the link between biological results, clinical condition and treatment adaptation. They also learn to monitor their diet and pay attention to their digestive problems. A technician familiarizes them with the e-tablet for half an hour and teaches them to fill in a simplified questionnaire. "In reality, the interface created by the operator is clear and fun enough that an elderly person who does not know computers or who sees badly can transmit monitoring data in less than 15 minutes", says Professor Zaoui, head of the nephrology department at Grenoble University Hospital, pilot of the experiment[16].

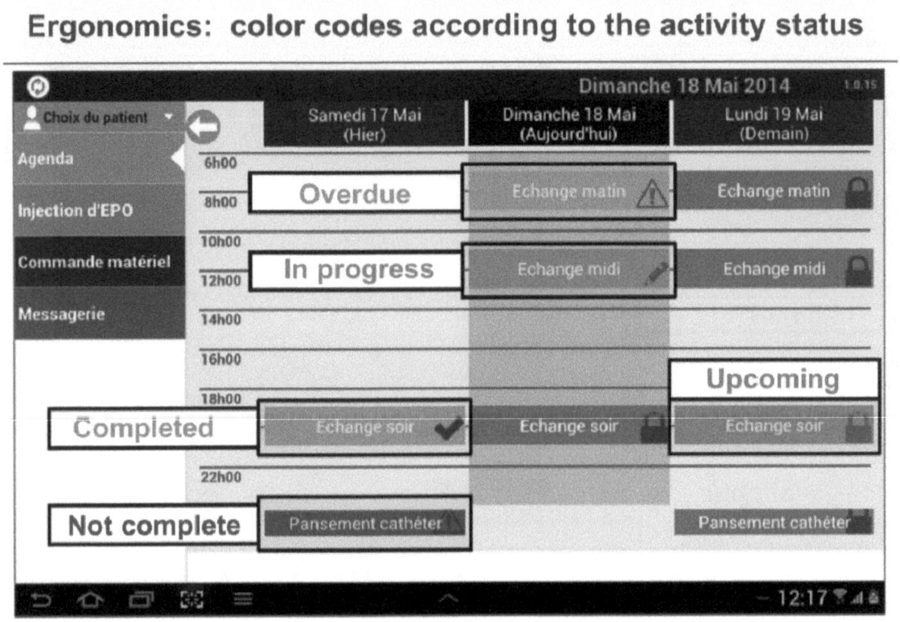


**Figure 5.6.** *Patient's electronic logbook in Ap'Telecare (Source: [DAB 14]. For a color version of this figure, see www.iste.co.uk/bardy/telemedicine.zip*

Patients now interact with the service via the e-tablet connected to the traditional telephone network. Every week, they take 15 min of their time to fill in monitoring parameters (blood pressure, weight, edema) online. The data entered at home is then encrypted by the system and automatically sent to the medical center via the operator's secure transmission network. They are hosted in complete confidentiality on approved secure servers. The healthcare professional who receives the information reads it in the form of histories, graphs, statistical reports and/or alerts

16 http://formaticsante.fr/index.php/dossiers/sante-soins-et-tic/69-telesante-telemedecine/629-le-telesuivi-a-domicile-des-patients-insuffisants-renaux. Page consulted on February 2, 2015.

triggered by the system in case of unusual parameters or indicating a risk for the patient (fatigue, shortness of breath, weight gain, biological markers that have reached an alert range). Depending on the patient's state of health, the doctor sends the patient therapeutic messages and advice. This e-health experience is based on support, follow-up and real-time exchange based on the definition of common therapeutic objectives and a therapeutic education project. This treatment has a triple objective: to stabilize as many people as possible without dialysis[17], to successfully maintain them at home during the dialysis stage, and when a transplant is necessary, to accelerate its completion. It is based on a system designed in partnership with TMM software, which has developed the Ap'Telecare application mentioned above. Note that this application allows data sharing on a local network, unlike the personalized medical record (PMR). The acknowledged failure of the PMR[18] in France, as well as in the United Kingdom and Germany ("Gesundhheitkarte"), paved the way for the development of electronic logbooks and a computer management system called "blockchain" (see definition below).

DEFINITION 5.2.– A *blockchain*, or chain of blocks, is a technology for storing and transmitting information without a control body. Technically, it is a distributed database whose information sent by users and internal links to the database are checked and grouped at regular time intervals into blocks, the whole being secured by cryptography, and thus forming a chain. By extension, a block string is a distributed database that manages a list of records protected against falsification or modification by storage nodes. A *blockchain* is therefore a distributed and secured register of all transactions carried out since the start of the distributed system[19].

### 5.3.3. *Follow-up action (dialysis patient)*

The Ap'Telecare application, which can be used on the smartphone or tablet (dedicated to the project or not) of the dialysis patient, also makes it possible to present the treatment path programmed in the back office. This application, available on Android and the App Store and accessible via all terminals with internet access,

17 As Dr. Agnès Caillette-Beaudoin, director of the Calydial dialysis center in Lyon, explains: "better management of patients, better monitoring of their pathology can delay the start of dialysis for up to three months. Three months of dialysis is 20,000 euros" [CAI 14].
18 See also on this point, "Dossier médical personnel, la bonne idée qui tourne au fiasco". Source: https://www.allodocteurs.fr/se-soigner/droits-et-demarches/dossier-medical/dossier-medical-personnel-la-bonne-idee-qui-tourne-au-fiasco_12498.html.
19 Source: https://fr.wikipedia.org/wiki/Blockchain#cite_note-3.

allows the setting of personalized care paths for dialysis patients. Important statements: the use of this application is subject to declaration. In addition, its deployment in an institution must respond to a joint dynamic of management, physicians and a paramedical team. A support system is strongly advised to allow an optimal use of the application and especially a parameter setting adapted to the practice of the patient suffering from chronic renal failure. This dialysis patient monitoring solution, co-developed by Calydial with the support of ASTRH@, a telemedicine association in Rhône-Alpes, was developed in an innovative partnership mode such as a Living Lab laboratory with the I-Care Rhône-Alpes Cluster. The patient interface (see Figures 5.5 and 5.6), oriented toward ergonomic use, very close to a peritoneal dialysis logbook, is very interactive. The data entry is simplified and adaptable, with the possibility of integrating chronology, well beyond the follow-up day (week, month) such that an electronic diary is possible.

The patient is offered and has programmed on their terminal their path as a peritoneal dialysis patient. For each patient, the notion of care activity is defined, namely peritoneal dialysis exchange, diabetes monitoring and validation of erythropoiesis stimulating agent injection (see Box 5.2)[20]. Thus, a true path of personalized care is created by the health professional in a transparent way. By their daily entries, adapted to their wishes or abilities, they will be able to validate the stage's course and transmit any information to the medical team. In addition to specific peritoneal dialysis care, pain assessment, pharmaceutical orders or any other activity that could be integrated into the patient treatment course, can be added. Ergonomics can be configured for each patient, according to their abilities and wishes. The uses for the professionals are totally customizable: definition of the values to follow and the alerts that can occur, and then creation of the activities regrouping them [CAI 14, p. 49].

> A first study, Télédiab, was conducted from 2007 to 2009. An average 0.9% decrease in HbA1c (glycated hemoglobin, which retrospectively measures the average blood glucose concentration over the past few months) was recorded over six months without increasing the risk of hypoglycemia and without increasing medical time. The study also showed patient interest in the device. At the end of the experiment, 70% of users wanted to keep Diabéo. The Diabéo application can only be used in the Telesage clinical study, after activation by a health professional authorized for this study.

**Box 5.2.** *Telediab and Telesage clinical studies*[21]

20 https://fr.wikipedia.org/wiki/Érythropoïèse. Page consulted on October 2, 2015.

21 Source: http://www.medecingeek.com/Diabéo-une-application-reussie-de-suivi-des-diabetiques/. Page consulted on December 3, 2014.

DEFINITION 5.3.– Erythropoiesis is the set of erythrocyte (red blood cell) production processes in red bone marrow from totipotent hematopoietic stem cells under the dependence of erythropoietin (EPO).

### 5.3.4. *A "chrono-responsible" patient*

The patient suffering from renal failure is a "supported participant" in their own health, in the sense that the act of autonomous peritoneal dialysis at home is part of a *therapeutic pathway* programmed according to a protocol determined by the nephrologist according to the indications provided by the telemonitoring and the reporting of daily biological data. The care, performed offline, is therefore closely monitored by the practitioner. Although operationally autonomous, unlike the cardiac patient discussed above, the dialysis patient is under directed therapeutic supervision.

The interest of telemonitoring due to the Ap'Telecare application therefore lies in the possibility of empowering the patient with regard to their day-to-day temporal management of the disease. The electronic logbook (see Figures 5.5 and 5.6) allows not only teleconsultation, but also televigilance, a true remote sentinel of the evolution of the disease and its management by the patient. Thus, the latter can precisely bring certain chrono-sensitive information to the attention of the telemedical team in their electronic logbook: injection times for EPO, exchange of peritoneal dialysis, replacement of dressings around the catheter, etc. It is indeed admitted in a hospital environment that a patient who no longer transmits, or ineffectively transmits, their data is a drifting patient, very often in refusal of their disease. Then they are in danger; hence the need to make the latter responsible for the temporal management of their disease, particularly through therapeutic education, to enable them to avoid acute imbalance, medical emergency, chronic metabolic imbalance or long-term complication risks. In a telemonitoring setting for patients with renal failure, the concepts of "accountability" and "shared temporality" of the disease seem inextricably linked. However, this juxtaposition needs to be clarified. On the one hand, it should be stressed that patient empowerment here is nothing other than the concretization at the individual level of the intrinsically preventive and delegative strategy of home telemonitoring. This is essential because the operational dynamics of telemonitoring, based on real-time data sharing, jointly aim to encourage greater patient empowerment and accountability in the years to come. In other words, patient empowerment at home is following an exponential trend that will apply, in the future, to all telemonitored diseases (chronic or not). The marketing of "empowering" medical devices, which is now taking place, will gradually relegate the more traditional "connected" or "communication" devices to the background.

On the other hand, the sharing of medical data, in real time, between telemonitoring stakeholders implies the tacit implementation of a shared responsibility. For the practitioner, it is a question of communicating to the patient a reliable diagnosis or recommendations, as soon as possible, on the basis of the data provided. For the patient, it is a matter of respecting the spirit and intent of the follow-up and medical care that has been delegated to them. The management of the temporality of the disease occupies a central place in the follow-up of the patient, and, contrary to the telemonitored cardiac person, engages the patient suffering from renal failure in a process of "chrono-responsibility", which must be explained here.

As previously indicated, acute chronic renal failure (chronic disease) is irreversible and therefore incurable. Its telemonitoring and the resulting telecare relationship is based on a principle of disease stabilization. Continuous monitoring is therefore necessary to maintain the patient in a satisfactory state of health enabling them to organize their social and family life. However, the patient's responsibility for chronic care, which therefore has a dual therapeutic and psychosocial objective, depends largely on its compliance with the *timing* imposed by telemonitoring and the sharing of responsibility for care. The benefits of time accountability are twofold. On the one hand, the patient becomes a full co-participant in the telecare relationship. On the other hand, the person benefits from the chronic proceduralization and protocolization of telecare that allows, on the basis of real-time data sharing, the potential risks of chronic disease to be dealt with.

If the shared temporality of follow-up and care between the patient and the practitioner thus facilitates the chrono-responsibility of the person in chronic renal failure, an increased synchronization between the individual and the communication device makes it possible, more generally, to redefine the concept of "patient" in the context of medical telemonitoring.

## 5.4. Diabetic patient as "synchronized interactive patient"

### 5.4.1. *A self-learning application*

The endocrino-diabetology team of the *Centre hospitalier sud-francilien* (CHSF, South Francilien Hospital Center) and CERITD (French center for diabetes treatment intensification studies and research) have developed a downloadable mobile phone application for patients with type 1 diabetes[22]. This application, known as the eponymous program Diabéo (see Chapter 4), facilitates the follow-up

22 The healthcare team and the research team at CERITD (*Centre d'études et de recherche pour l'intensification du diabète*, Corbeil-Essonnes), led by Dr. Charpentier (*Centre hospitalier sud francilien*), created the first "user-friendly" electronic logbook in collaboration with software publisher Voluntis.

of their treatment and alerts the medical team in case of irregularity. Diabéo has also been designed to help patients manage their disease on a daily basis. It offers an aid to the calculation in real time of slow and fast insulin doses, depending on diet and physical activity, as prescribed by the doctor. The application is self-learning: if a patient is hypoglycemic or hyperglycemic after meals, Diabéo proposes to increase or decrease their doses. Diabéo's goal is also to strengthen the doctor–patient relationship through automatic transmission of results. An automatic analysis tool is integrated: it alerts the professional in the event of too frequent anomalies (hypo- or hyperglycemia, for example). About 20 different alert messages are defined. Diabéo was the first communication device in France to be combined with a telemonitoring service for patients undergoing complex insulin therapy. Diabéo consists of a communication device used by the patient (see Figure 5.7). This is a smartphone application with real-time titration help. The application is equipped with a self-learning medical algorithm. Diabéo offers a telemonitoring service: the doctor has a monitoring dashboard with automatic analyses and the nurse has access to a protocolized platform for telemonitoring.

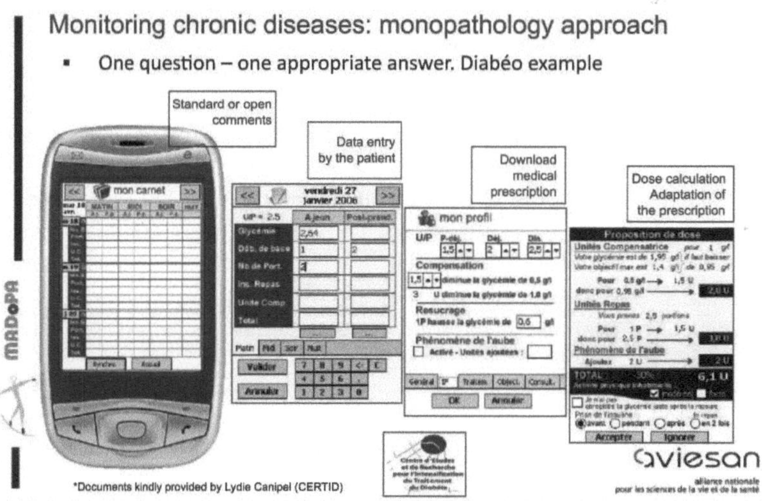


**Figure 5.7.** *Diabéo communication device used by diabetic patients (Source: [DUC 14])*

The demonstrated benefits of the Diabéo solution are medical (better glycemic control without increasing hypoglycemic risk), economic (reduction in transport costs and number of trips) and organizational (gain in medical productivity, better control without increasing medical time)[23].

23 See the Telediab 1, study (A. Penfomis, 2009, 180 type 1 diabetes patients, six months). See also Box 5.3.

For the professional, Diabéo provides 24-hour continuous access to all the patient's blood glucose data. The patient tests their blood sugar about seven to eight times a day, before and after each meal, and depending on the person's state of health. The blood glucose reading is done as follows: the patient places a drop of blood on a test strip, then the glucose level is displayed directly on the smartphone display (an electronic meter is connected to the smartphone). The phone records all blood glucose levels for better monitoring. The patient manages the insulin injection adjustments themselves from the data displayed on the screen. Insulin, which the pancreas no longer produces, is delivered by a small pump that regulates glucose. The patient enters the glucose level; the Diabéo application installed on the smartphone is able to automatically calculate the correct insulin dose. In summary, Diabéo is the physician's electronic prescription, in the patient's pocket, that gives the patient the exact dose of insulin. They are a patient who feels close to their doctor and reassured, who sees, from the right dose of insulin calculated by their electronic logbook, their blood glucose results improve, as proved by the results of the multicenter study Telediab1.

Seventeen French diabetes centers followed 180 adults with type 1 diabetes who were treated by basal bolus (more than six months per pump or multiple injections) and chronically unbalanced (HbA1c > 8% on at least two consecutive occasions and at least initial HbA1c = 9.1+ or –1.1%) for 6 months. These 180 patients were randomized into three groups: a control group benefiting from the classically recognized follow-up with a quarterly face-to-face consultation, a group with the electronic logbook and quarterly face-to-face consultations, and a group benefiting from both the electronic logbook and short telephone medical consultations of 5 min on average without quarterly face-to-face consultations. After six months, HbA1c was improved by 0.9%, without any increase in hypoglycemia or medical time devoted to these patients. The patients did not have to take time off work or travel to the medical office.

**Box 5.3.** *Results of the Telediab 1 clinical study (Source: [DUC 14])*

### 5.4.2. *The telepatient as an "interactive" patient*[24]

The increased synchronization between the telediabetology system and the individual, through a self-learning and empowering device, gives the individual the status of "interactive" patient: the individual is no longer (simply) a patient data producer (cardiac telepatient), or a responsible co-participant under the direct control

---

24 In a broad sense, this neologism is used to speak of the patient as a participant in their own health, as a "health participant" (*santacteur*; see Chapter 1, section 1.2).

of a practitioner (renal failure telemonitoring), but a "committed" patient sharing with an automated system the responsibility of telecare in real time, under the *indirect* control of a practitioner. As a participant in their own health, empowered by an application equipped with a self-learning medical algorithm, the diabetic patient is thus jointly led to share the *decision* and *practice* of care with the empowering device – even in a country where studies show chronic patient dissatisfaction in decision-making.

As explained by cardiologist Jean-François Thébaut, a member of the *Haute autorité de santé* (HAS, High Authority for Health):

> *In France, we are the first country to have legislation in place on patient information, the law of March 4, 2002. And yet we are the country where patients feel the least informed. Let's talk about the Commonwealth fund. It is an international organization in which health insurance and the* Haute autorité de santé *participate for France, which carries out surveys in 11 international countries. It does the same survey in all 11 countries and repeats them regularly, 2008, 2011, and now 2014. It asks the same questions in the countries. This allows countries to be ranked according to patients' feelings. These questions are asked to patients one year, and to professionals the next. France is among the 11 countries, the worst placed for feeling – I am not saying that it is accuracy – but it is for feeling the information that patients have had after a consultation, whether for the general practitioner or the specialist doctor. We are the country where patients feel the least involved in decisions [...]. Other surveys have shown in the same state of affairs, the divergence of perceptions. There was a survey done in 2011 by INPES (*Institut national de prévention et d'éducation pour la santé*) [National Institute for Prevention and Health Education] on diabetes, where we see the differences in perception that there are in the decision that may have been made between the patient and the doctor. In almost 60% of cases, the patient clearly has the impression that it was the doctor who made the decision, whereas the doctor had the impression that he or she only made the decision in about 20% of cases, that they shared it in the other cases and in almost 25% of cases that it was the patient's decision*[25].

25 [THÉ 15] http://www.education-preventionsante.fr/wp-content/uploads/2015/01/Dr-Jean-François-Thebaut.pdf. Page consulted on January 20, 2015.

Because of their mastery of the device (electronic logbook) and of the possibility which is given to them to carry out their own care at home (injections of insulin), without direct intervention of the practitioner, the interactive patient acquires, beyond an increased autonomy toward the medical team, in-depth knowledge of their disease and the level of care that corresponds to them because of the automated system (rate of insulin required compared to the rate of glycemia noted). However, this opportunity can only be offered to diabetic patients if they strictly follow the telemonitoring protocol at home and if they jointly adopt its underlying values and adapt, moreover, to their rhythmicity.

## 5.5. Compliance and concordance

From a techno-medical point of view, the management of the temporality of the telemonitored patient referred to in the previous pages, all chronic diseases combined, implies to varying degrees strict patient compliance with the procedures implemented as part of their home telemonitoring[26]. What is targeted through the principle of compliance is patient adherence. This principle of compliance is concretized by the negotiated implementation of "disciplinary" behavior at the patient level. Self-management of illness and care, at the pace of daily readings and data transfer of life and independent care, is the key element responsible for the telemonitoring protocol. However, this principle of compliance does not prevail solely as a *modus operandi* for home telemonitoring: it is also a principle that is certainly empowering but it is also binding, a sort of *modus vivendi* around which the patient–physician relationship, itself subject to time constraints (schedule of visits, monthly monitoring appointments, etc.), is structured.

To this must be added the need to coordinate and connect the players using the connected objects, which aims to constantly update, if possible in real time, the life data and the notable actions carried out, by the patient. The idea is to create a kind of "information node" from which all participants can enter into dialog and act jointly with the goal of well-being. This alignment of the knowledge of all participants on the patient favors a concordance of practices in order to achieve the best possible result for the patient. What role does the cardiac patient play in this process?

The *cardiac patient's* compliance with the principle of real-time management by proxy of their illness, agreed to by the telemedical team in charge of their follow-up, presupposes acceptance of the value of medical authority and expertise. The strict control of the evolution of the disease by the practitioner – from automated reception and analysis of data to diagnosis – excludes the patient *a priori* from the

26 See also the definition of compliance (Chapter 2, Definition 2.1).

decision-making loop. The "management" of the "organ experience" (the heart) measured by the object (the defibrillator) is only shared *a posteriori* with the patient during teleconsultation. Placed *de facto* in a *passive situation* and unable to access, in real time, the data likely to remove the unpredictability of their disease, the patient seems to have no other choice than to rely on the medical authority holding a knowledge based on an "already measured experience" of their disease. Under these conditions, combined with the principle of (patient) compliance with the technical means implemented within the framework of telemonitoring, mentioned above, which is not the only measurable indicator of the patient–physician relationship in the context of telemonitoring[27], the value of (medical authority) seems to prevail over patient autonomy. More importantly, the principle of patient compliance that results from the value of authority inscribed in the dynamics of real-time monitoring also seems to have the effect of excluding from the decision-making loop the patient's bodily experience, often considered as profane knowledge with no real scientific value. It is certainly the responsibility of social caregivers to pass on to the practitioner, including as part of the telemonitoring protocol, the patient's subjective knowledge of their state of health (personal feelings, bodily experience, psychological state), in particular for the treatment of chronic renal failure. As previously indicated, however, this information seems to have only relative value in the telemedical management of chronic disease today, with some exceptions[28]. On this point, the concept of "*person*-centered care" proposed by Claudia Bartz – for her it is the concrete person mentioned above and not the abstract concept of patient[29] – is particularly illuminating and useful in recharacterizing the value of the individual's subjective experience and knowledge in the telecare relationship.

Note that in the field of digital health many states are on the authorization regime. This is, for example, obtaining the prior consent of the cardiac patient before implanting a pacemaker. While it is important to authorize a participant, it is no less important to certify a treatment. Any protocol of care, delegated to the patient or not, requires a guarantee that passes by the certification of the care. The review of the principle of conformity, mentioned earlier, does not call into question the modalities or the relevance of certification, but places more emphasis on the *tacit* agreement between the participants (patient–physician) in order to rigorously observe the care practices delegated and implemented at a distance.

27 Other factors that have a strong impact on the patient–physician relationship and its temporality are the empathy of the practitioner and caregivers or the harmonization of caregiver–patient temporalities (see Chapters 5 and 6).

28 See, in particular, [BAR 10]. See also [MOR 12] for a comparative analysis of this concept in the literature. For a more in-depth analysis of this concept, see [BAR 14b].

29 See also section 2.3, "The person in telecare".

*The person with renal failure*, empowered through their home dialysis system, or the *diabetic patient* using a computerized insulin pump and synchronized to a self-learning device, are subject, like the cardiac patient, to the same principle of compliance with the technique. Unlike cardiac telemonitoring, however, home dialysis telemonitoring or telediabetology systems aim to actively involve the person in the practice of home care. The principle of patient compliance is based here on a disciplinarization logic that presupposes adherence to two distinct operational modes of telemonitoring: *time-sharing* and *synchronized mode*. What are the main values inherent in this "temporal discipline" of the patient?

It should be noted that the autonomy of the dialysis patient or diabetic is conditional: it implies that the person keeps their dialysis / electronic blood glucose monitoring logbook up to date, due to which the "treatment course monitoring" is possible. The smooth progress of the telemonitoring protocol depends on the patient's dialysis and diabetes data being reported and shared with the practitioner on a daily basis. The chrono-responsibility of the patient is thus one of the main modes of their discipline. In this sense, the daily sharing of biological data confers a binding, but necessary, prescriptive value to telemonitoring. This constraint further reinforces the medical authority exercised over the patient (dialysis patient or diabetic) because the latter, despite being a co-participant of the care, is in no way co-deciding *with the practitioner*: Their knowledge about the disease or their "competence" is not taken into account, nor formally recognized, in the telemonitoring protocol. The ability of the chrono-responsible patient or the synchronized patient to act is therefore limited to the self-administration of care within the limits of a *pastoral* approach to telemonitoring, relayed by self-learning computerized systems. This observation, shared by many telemedicine observers today, has recently led some researchers to propose new models for decision sharing. For example, the three-step model, proposed by English researcher Glyn Elwyn, (1) gives the patient a choice, (2) describes the possibilities of care (possibly using decision support tools) and (3) *helps the patient in their reflection and decision-making* [ELW 12]. This model is partially implemented in the negotiated telemonitoring of chronic patients. However, current telemonitoring practices, particularly those described in this chapter, are still mainly focused on the analysis of quantifiable disease metrics that do not integrate certain subjective disease factors (such as the patient's body experience and values); the therapeutic value and epistemological scope of these are nevertheless measured in Chapter 6.

Finally, in addition to the principle of patient compliance with the multiple rhythmicity of telemonitoring, which reveals the medical institution's view of the patient and their temporality, there is the primacy of concordance between all those involved in the telecare relationship. Indeed, the efficiency of telemonitoring presupposes flawless coordination, both in technological and human terms, which makes it possible to ensure the meshing of the multiple temporalities of telemonitoring. The principles of conformity and concordance, thus understood, are at the basis of what could be called a "temporal economy" of telemonitoring through which the specificities of the contemporary telemedical view of chronic disease are ultimately drawn.

## 5.6. Harmonizations and phase shifts

The time-saving of home telemonitoring, based on the principles of compliance and concordance, is mainly based on three modes of time-sensitive management of the disease – by proxy, by chrono-responsibility and by empowerment – which are carriers of values whose examination makes it possible to enlighten the telemedical perspective on the patient today. Thus, the value of power of attorney is fundamentally pastoral, in the sense of Nikolas Rose[30]. Because of their control of the operational methods of telemonitoring, the practitioner imposes themself as an authority figure holding knowledge about the disease, which is constantly updated, to which *the cardiac patient* has no immediate access. In this context, their typical profile is both that of a "patient guided" by the practitioner, and that of a "patient out of step" with the protocol. Delayed cardiac patient update, at the teleconsultation stage, is a late harmonization practice.

The chrono-responsibility of *dialyzed telepatient* care is a very particular method of discipline, both useful and constraining, which presupposes the negotiated caregiver–care receiver interactions of moral feelings (trust, respect) and specific home care practices based on the shared expertise of the practitioner. As in the case of telemonitoring of the cardiac patient, the practitioner exercises a pastoral responsibility, an essential condition for the proper conduct of home care, but by delegating to the patient a certain number of tasks according to a programmed chronology. The value of co-responsibility in the management of telemonitoring operational modes is, therefore, the correlate of the chrono-responsibility of the dialysis patient on a daily basis. Like the value of power of attorney, mentioned above, the value of co-responsibility nevertheless constitutes the principle of compliance, which tends to relegate the patient's knowledge or subjectivity to the margins of the telemonitoring protocol. A similar, but perhaps more radical,

30 Following on from his analysis of the power of contemporary medical pastoralism, Nikolas Rose draws the following conclusions: "We are not talking here about pastoralism in which the shepherd who knows his flock well shows the way to his sheep lost in doubt. This pastoralism implies dynamic cause–effect links between those who advise and those who are advised. These new health priests espouse the ethical principles of informed consent, autonomy, volunteer action, choice and non-directive counselling. In the age of biological prudence, these ethical principles inevitably translate into the emergence of micro-technological systems for the management of communication and information, which inevitably have a normative and directive vocation" [ROS 07, p. 29].

conception of coresponsibility also seems to be emerging through the empowerment of the interactive patient (diabetic patient) by intermediary object. Indeed, the responsibility for the care is validated by the automated system that transmits in real time the recommended dose of insulin to the patient. The practitioner is certainly not absent from the loop since the application gives them access to blood glucose readings. However, this automated validation is a "systemic" mode of protocol compliance.

An examination of the values underlying the time-sensitive management methods of the disease sheds light on the specific features of the harmonization logic specific to telehomecare. This logic is based on an imperative of compliance for therapeutic efficiency purposes without however including the patient in the decision-making loop. It also results in a permanent and "prudential" telemonitoring of the person, more or less sensitive to their pace of life, based on a quantitative conception of the disease and care. Finally, this logic contributes to blurring the boundaries not only between the patient's experience and the experience measured by the empowering objects, but also between the value systems and the modes of representation of the participants in telecare. However, the logic of harmonizing telemonitoring practices and times cannot mask a certain number of discrepancies between the caregiver's concepts of the disease and that of the patient, between the disease observed and the disease experienced, between the evolutionary internalization of the person's perspective and the externalization of the telemedical perspective centered on the illness in its clinical intimacy.

Thus, the telemonitored chronic patient may sometimes be led to experience a situation of psychic individuation, in the Simondonian sense of the term[31]. The person may become out of phase with themselves and their illness as a result of the dynamics of telemonitoring. On the one hand, this phase shift can induce a potential withdrawal into oneself and the possibility of a cleavage for the patient, who is likely to develop a "double" awareness of their disease, inside *and* outside of oneself, but it can also generate antagonism between individual *and* technical perceptions of the disease. Affection, far from being an object of pure quantification *or* an inseparable experience of the patient, is a complex, ambivalent reality in the context of telemonitoring, which notably calls for a better understanding of the person's "inner time" as well as a rapprochement, if not reconciliation, of the patient's "subjective time" and the "objective time" of the telemonitoring device.

31 Here, we take up and adapt the principle of psychological individuation defined by Simondon in *L'individuation à la lumière des notions de forme et d'information*, for example in the following statements: "There is a psyche when the living does not fully materialize, preserves an internal duality" when it cannot get to "be completely appeased and satisfied in itself" when "affectivity is overwhelmed" [SIM 05, p. 165].

This approximation seems possible through mediation by the instigators of telemonitoring projects sensitive to the multiple temporality of the patient. The implementation of the conditions of this mediation seems all the more urgent today as the so-called "autonomized" and "empowered" patient gradually acquires the mastery of "new technologies", a kind of digital version of the self-study techniques analyzed by Michel Foucault, which lead them to replace the practitioner. To achieve this mediation, a holistic approach to the temporality of the patient, centered not only on the chronology of their illness but also on their body experience, appears to be an interesting avenue for reflection.

The corporeal and experiential existence in fact refers to a subjective perception of time that finds its foundation in the three classical dimensions of time:

1) time as duration ("Chronos") – for example, the use and distribution of daily tasks (work, care, leisure);

2) time as the duration of life ("Aiôn") – for example, the rhythm of daily life;

3) time as an opportune moment ("Kairos") – the time of interaction with devices and what they tell us at the moment T of our present and future[32].

The underlying epistemological value of these concepts of time refers to what could be called an "ontology of time" of telecare: both long time (that of illness), short time (that of daily life) and repetitive or cyclical time (that of care). These are three temporalities that the person goes through together and that constitute the person's experience. But if the patient's experience in the telecare context has, on the one hand, a major role to play in understanding the evolution of the disease and finds its ontological basis, on the other hand, in the combined concepts of Chronos, Kairos and Aiôn, its management also has a *therapeutic value* (see Definition 5.4).

DEFINITION 5.4.– The *therapeutic value* in a telecare setting refers to the recognized benefit of a device, a care practice or experiential existence that optimizes disease and patient management.

By placing the emphasis on the therapeutic value of the recognition of experiential existence, the analysis of telecare ultimately points to the patient's "capability". According to Amartya Sen's definition, capability refers to an individual's actual ability to choose various combinations of functioning, in other words to test the freedom that they actually enjoy[33]. In the context of telecare, this

32 On this point, see also, Chapter 7, and in particular Box 7.3.

33 The approach to capabilities has subsequently undergone important developments, notably in the work of the philosopher Martha Nussbaum, who links the approach to capabilities and the policy of care proposed by Joan Tronto, and underlines the anchoring of this approach in

freedom, for the patient, is that of sharing experience in order to optimize care. The possibility of expressing this freedom can also contribute, on the one hand, to remedying the discrepancies imposed by telemonitoring protocols and, on the other hand, to reclassifying the status of "chronic patient" or "telepatient": no longer as an "object" but as a "subject" of care whose experience is part of multiple temporalities and makes sense in therapeutic management.

What lessons can be learned from examining the chrono-diversity of home telemonitoring, which underlies chronic patient care and its temporality regime and illuminates the contemporary telemedical perspective?

First, in order for the patient's temporality experience in a telemonitoring context to be possible as a form of knowledge, it is necessary to give meaning to the sick body in its digital reconfiguration, to examine the potential effects of the dynamics of telemonitoring on the person and to elucidate its status in light of the reexamination of the principles of conformity and concordance specific to telemonitoring.

Second, just as the new medical spirit based on observation and its methods, initiated by Bichat according to Michel Foucault, was "nothing other than an epistemological reorganization of the disease where the limits of the visible and the invisible follow a new design" [FOU 09, pp. 269–270], the contemporary telemedical spirit seems to be nothing more than an epistemological reorganization of care practices. The management of chronic disease in the context of home telemonitoring takes concrete form in particular through the development of two types of knowledge. On the one hand, the knowledge built by the patient (or biological knowledge of oneself) and, on the other hand, the knowledge acquired through telemonitoring (or biomedical knowledge about the individual being telemonitored)[34]. However, this knowledge is not used in equivalent terms. Thus, the consideration of subjective knowledge about chronic disease, generally delegated to social or family caregivers, still appears marginal where it might be useful, including on a therapeutic level, to give a more central place to lay knowledge about the disease and its temporality in the telemonitoring protocol.

The examination of the chrono-diversity of telemonitoring devices and participants, and more particularly the examination of the *temporal figures* of the chronic patient being telemonitored, make it possible to clarify the specificities of the temporal distribution particular to the home telemonitoring of chronic disease by

the tradition of liberal thought at the same time as its Marxist inspiration. For Amartya Sen, the notion takes root in the theory of social choice as well as in the moral philosophy and in the philosophy of analytical action (see [SEN 87]).

34 See Chapter 3, section 3.5.

emphasizing in particular the therapeutic interest of a resolutely holistic approach to care that presupposes jointly taking into account the physical and psychological experience of the patient. This double observation thus makes it possible, in the latter instance, to highlight the major ethical challenge facing telemedicine today: how to reconcile medical technicality with human care.

# 6

# Experiential Knowledge of the "Subject of Care"

*Against the excesses of a technology often deemed inhumane, running counter to a division of labor which has led some to lament a possible "clinical death", here is the care that is now in need of consideration. It is not only about homage that is given back to the commitment of staff and casual mention of small gestures that humbly alleviate the patient's suffering. What is at stake is the rediscovery of the anthropological basis on which medicine – all medicine – merges its humanity and its effectiveness.*

Dominique Lecourt

QUESTION 6.1.– How does the person live through their illness in the context of care and remote monitoring?

Answering this question involves expanding on the person's feelings in order to understand the patient. The person undergoing telemonitoring is actually irreducible to temporal profiles that are constructed in telecare practices and algorithmic "placing of numbers", which take place because of new digital health technologies. After the review of medico-economic evaluations of telemedicine and case studies, the analysis of the role of temporality and emotions in the construction of the patient experience (Chapter 3) and, more generally, of the temporal distribution between telecare actors (Chapter 4), we propose to make the voice of people engaged in a protocol of autonomous care and telemonitoring at home be heard.

This chapter is thus intended to identify the physical and psycho-emotional factors involved in the self-management of illness at home and in the phenomena of stress inherent in telecare. To ensure a quality of life, it is in fact necessary that the patient is able to take action on sources of stress to prevent imbalances and escape emotional "overflow". To understand the mechanisms underlying the patient's well-being is, however, not an end in itself. It is, therefore, suggested that we do not dwell here on the results established at this level by patients nor stick to cataloguing the benefits and flaws of telemonitoring systems, but show how the extended care of temporal and psycho-emotional aspects of the disease, essential to the construction of the *care subject*, is one of the ethical priorities of telecare today.

The distance and information dimension of telecare models presented in Chapters 3 and 4, combining advanced technology and medical expertise, seem to focus on automation and self-management of care devices. Heir to an undermined biomedical model[1], telecare is however evolving jointly toward a new model of *relational medicine*, sought by a growing number of patients and practitioners today. But this development, in the view of interviewees, is too slow. The reoccurring testimonies highlight a need for empathetic listening and psychological counseling. This is probably due to a lack of human and financial resources. In their official report, the *Société Francophone du Diabète* [Francophone Diabetes Society] take a similar finding with a novel proposal: "today, the time spent by the doctor listening, educating and advising his patient is not compensated as such nor reimbursed by social security. The introduction of a flat fee by pathology would take into account this time spent with the patient and stop the inflationary logic of the fee"[2]. But this situation is also a result of a misunderstanding of the patient undergoing "telecare" which the present chapter wishes to address.

Interviewee testimonies allow a behind-the-scenes view of the telemonitoring of chronic diseases at home and enable us to assess the emergence conditions of a new therapeutic alliance valuing increased care in terms of the everyday perception of the disease and telemonitoring expressed by the patients. To identify the specificities and the epistemological value of this new alliance, it is proposed to present, beforehand, the methodological approach adopted in this chapter as well as the various actors who have helped to bring this research to a successful conclusion.

## 6.1. Constructing links with patients

Telemedicine is a sensitive area. Patient associations and federations, as well as some organizations and healthcare professionals, may be reluctant to share contact details and the information they have on patients and their experiences of chronic disease and telemonitoring. The recommended procedure, if not required, to complete a research project on contemporary telemedicine is, in principle, the following:

1 That of medicine focused on the biological order of the patient. On this point, see Chapter 5.
2 http://www.afd.asso.fr/actualites/le-livre-blanc-du-diabete-de-la-societe-francophone-du-diabete-001357. Page consulted on March 3, 2016.

– presentation of the project to the researcher's university ethics committee;

– validation of the project and the chosen research method;

– contact established with associations or patient federations, organizations, establishments or health professionals;

– submission of the research project to the actors;

– validation of the project by the competent authorities: association president, healthcare professional, ethics committee, general assembly, etc.

– linking and interviews with patients.

However, it was not possible to follow this model because the ethics committee of the university was not yet constituted at the time of research and did not allow us to bring the project to the actors' attention in the conventional way. Another way was: dialogue with patient associations and healthcare professionals involved in the telemonitoring flagship projects, eager to share and publicize their experiences, including at the academic research level. Methodologically, the format of an individual telephone interview with voluntary chronic patients has been chosen. Confidential and anonymized, this telephone interview was the most suitable formula to interview homecare patients, sometimes with limited mobility and availability, spread over the entire French territory. Along with this approach, the study was enriched by many testimonies from patients and patient experts engaged in actions of information and dialogue with their peers and, more generally, the public. Thus, the following resources could be useful to understand the daily life of the sick person:

– reviews of patient interventions (experts);

– transcription of patient videos (experts);

– testimonies of patients collected in various studies[3];

– healthcare professionals' testimonies on patient experience.

### 6.1.1. *Representativeness of the panel*

How to build a representative patient panel? According to what criteria? With what objectives? First, one must bear in mind the purpose of the study: collecting testimonies from patients eager to communicate the psychological dimension of the disease and telemonitoring. Therefore, its original purpose was not to assess the quality of life of a population for statistical purposes across the information obtained

3 See References.

with patients, healthcare and social/family caregivers affected by chronic disease. Such a study would have required, on the one hand, the implementation of a research protocol with many care centers, hospitals and professional associations that the exploratory nature of the project did not justify and, on the other hand, the development of a psychometric evaluation tool.

Psychometric evaluation tools of quality of life and psycho-emotional factors of the disease are complex. They generally consist of scales or questionnaires, they realize a more restrictive quality of life evaluation, but they allow a standardized measure. At the individual level, the psychometric tool must take the person as its own reference, explore and analyze all the areas that make up the quality of life related to health, and weigh each of the areas against the other (see Figure 3.3), according to their importance for each individual. Few tools meet these characteristics. We can cite here the SEIQoL (Schedule for the Evaluation of Individual Quality of Life)[4], PGI (Patient-Generated Index)[5] and SQLP (Subjective Quality of Life Profile)[6].

Seeking patient associations and/or health research professionals for research also demands an investment on their part, agreed on a voluntary basis, in addition to their activities. This investment, all the more so since the number of interviewees was high, took different forms: reading the proposed research; discussing its relevance with the leaders of the association or members of the service; submitting, as appropriate, the said project for the approval of a general meeting or a committee; getting in touch with patients likely to participate in the survey; informing them of the project; communicating their contact information to the researcher; follow-up interviews; etc. Given the constraints imposed by the survey on patient associations, it was necessary, with upstream research, to limit the number of interviews and take into account the voluntary character of their participation, adjusting the format of the interviews to people in accordance with their experience, their desire to communicate and their sensitivity. Finally, it should be noted here that telemonitoring of chronic diseases at home, although a national priority defined as part of the 2009 HPST law, concerns relatively few patients in France (all diseases combined), as indicated in Table 6.1.

4 Concerning this tool and the activities of the evaluation workshop in prevention and health advancement, see: http://www.atelierdelevaluation.fr/page-30/outil-etp-education-therapeutique-du-patient-seiqol-mesure-de-la-qualite-de-vie-percue.html.

5 For a description of this tool, see: https://www.health.belgium.be/fr/patient-generated-index.

6 For a description of this tool, see: https://eprovide.mapi-trust.org/instruments/subjective-quality-of-life-profile.

| | Heart failure | Renal Failure | Diabetes |
|---|---|---|---|
| Total number of patients | 500,000[7] | 42,000 | 3 million[8] 8% (T1) and 92% (T2) |
| Telemonitored patients | 40,000[9] (treatment with defibrillator) | 250[10] (by independent daily home dialysis) | 15,000[11] (treatment by insulin pump) |

**Table 6.1.** *Estimates of the number of chronic telemonitored patients in France in 2016*

Although growing steadily, the percentage of telemonitored patients remains relatively low (around 8–10%) compared to the estimates of the total number of patients with chronic diseases in France. Several reasons may explain this percentage. To treat chronic heart failure, there are, in particular, several alternatives to the implantation of a defibrillator: non-pharmacological treatment (patient education, weight control, physical activity and nutrition regulation, control of cardiovascular risk factors), drug therapy (beta blockers) and surgery (heart transplant)[12]. The small number of telefollowed patients undergoing independent home dialysis can be explained by a lack of information about existing solutions of home care, by the fact that free healthcare may lead some older patients with a number of comorbidities, more significant at the time of their care in chronic

7 http://www.coeur-recherche.fr/insuffisance_cardiaque. Page consulted on September 1, 2015.

8 Health Monitoring Institute [Institut de Veille Sanitaire (INVS)]. http://www.InVS.Sante.fr/dossiers-thematiques/maladies-chroniques-et-traumatismes/Diabete/Generalites-et-CHIFFRES-CLES/Le-diabete-en-quelques-Chiffres-et-faits. Page consulted on November 12, 2015.

9 http://docplayer.fr/7949240-Table-ronde-1-telesurveillance-de-patients-atteints-d-insuffisance-cardiaque.html. Page consulted on November 12, 2015.

10 http://www.allodocteurs.fr/maladies/prostate-reins-vessie/insuffisance-renale-dialyse/la-dialyse-a-domicile-une-technique-a-developper_14749.html. Page consulted on November 12, 2015.

11 http://ceritd.fr/activites/soins-ambulatoires/. Page consulted on November 12, 2015.

12 Beta blockers are used in cardiology. They are drugs that block the action of the mediators of the adrenergic system such as adrenaline. Their intended use is to lower blood pressure and are indicated, in particular, in the management of heart failure. Their principal mode of action is based on the decrease in the activity of the nervous system. After a period of latency, they reduce hypertension by complex mechanisms: inhibition of renin secretion and decrease in cardiac output. http://www.Futura-Sciences.com/magazines/Sante/infos/DICO/d/Medecine-beta-bloquant-3060/. Page consulted on November 1, 2015.

dialysis, to require maximum assistance, and by the difficulty of acquiring autonomy in the absence of family support. Finally, the use of remote monitoring is not necessarily justified for type 2 diabetic patients, who account for 90% of the diabetic population in France. Indeed, the recommended treatments are usually based on respect for certain basic principles (an appropriate diet, the adoption of an active lifestyle, monitoring of capillary glycemia) and, where appropriate, on drug observance. Note the transition of increased development in recent years of so-called non-drug interventions to these audiences (see Box 6.1).

Non-drug and non-surgical interventions belong to three major sectors of intervention research: kinesiology, nutrition and psycho-education. These interventions are complementary to conventional drug and surgical therapies. They are only very rarely alternatives to these therapies in the state of current knowledge. They often involve active participation on the part of their users, by the adoption of a new health behavior. They have a more direct and faster action on health indicators (a few days to a few months) than solutions within a wider area: habitat management, transformation of the environment, social reorganization or new economic orientations.

Non-drug interventions are intended for the sick or those at risk of becoming so very soon. The first category, and the most important, is chronic patients (a notion opposed to acute diseases that require emergency care whose action on state health is almost immediate). These people represent about one-third of the general population, 20 million in France. Surgical or medical treatments have failed to achieve a complete cure of these diseases having become chronic. They represent 77% of diseases, of which the most significant are cardiovascular, cerebrovascular, respiratory, metabolic and cancer.

**Box 6.1.** *Definition and audience of non-drug interventions [NIN 13]*

On a quantitative level, the spectrum of the present research has proven to be relatively narrow from the outset. As a result, the chosen target for the study was therefore to identify *trends* in the perception of telecare expressed by patients by taking into account particularly the temporality of care (of people, objects and remote monitoring systems deployed) and the physical and psychological experience of the chronic disease.

In this context, the representativeness of the panel is due to two essential factors:

– the novel value of patient knowledge acquired from the experiences of the people interviewed as part of a semidirected confidential interview;[13]

13 The psychological interview has a qualitative vocation. It enables an overall assessment of the quality of life and psycho-temporal impacts of telecare. It is therefore not a standardized or standardizable assessment tool; it cannot be used in comparative studies or applied on a large scale.

– the trend value of this knowledge: patient association members, interviewees have also reported their shared experience of the disease, care devices and telemonitoring with their peers.

The primacy of a qualitative approach of an individual experience to the trend value, concerning the physical, temporal and psycho-emotional impacts (positive and negative) of telecare, characterizes the survey carried out with chronic telemonitored patients. It remains to be seen to what extent this approach allows us to develop a relevant reflection on the patient's status as a care subject and on the possible conditions of the spreading of good telecare practices. This issue will be addressed in a second analysis in this chapter.

### 6.1.2. *Preparation and scope of the interviews*

Seven steps marked the survey work conducted on chronic patients engaged in remote home monitoring protocols:

– drafting and definition of objectives;

– review of secondary sources (surveys, international studies, etc.);

– preparation of semidirected interviews;

– drafting of the progression of interviews and a model questionnaire;

– researching patients (experts) and making contact;

– preliminary interview with patients:

- *presentation of the research and explanation of the progression of the interview;*

- *recalling the objectives of the interview and the terms of confidentiality;*

- *telephone appointment;*

– semi-directed interview (40–60 mins) on the basis of a model questionnaire;

– synthesis and analysis of the interviews.

### 6.1.3. *Presentation of the project to the patients*

1) Preliminary interview

The purpose of this meeting was to present the specificities and research objectives to the patients, as well as the terms of the semi-directed interview. It was important to stress, at this stage, the purpose of the study: constructively supporting the current ethical reflection about the means to be implemented to improve the quality of life of the telemonitored person.

2) Semi-directed interview

This interview had three objectives: identify the impacts of telemonitoring, including temporal and emotional factors of the disease; identify the management of these factors by the patient as part of their protocol; understand how their telemedical care (if any) works. Based on a model questionnaire, each confidential and anonymized telephone interview, an average of 60 mins, covered 20 subjects (see Table 6.2), addressing the following four themes:

– the terms and conditions of the protocol of telemonitoring at home;

– the uses of care devices, monitoring and remote telemonitoring;

– the temporal and emotional disease factors and their care;

– the qualitative assessment of the telemonitored person by the patients.

| Themes of the interviews |
|---|
| Specificities and practices of care devices |
| Specificities and uses of the communicating objects |
| Experience of the illness |
| Stress and emotions |
| Specificities of the chronic disease |
| Sense of the future |
| Doctor–patient relationship/caregiver |
| Daily time use |
| Pace of life |
| Perception of time |
| Orientation in time |
| Biological and circadian rhythms |
| Quality of life |
| Experience of the illness through objects |
| Treatment and observance calendar |
| Frequency and duration of follow-up activities |
| Doctor–patient relationship/caregiver |
| Diet |

**Table 6.2.** *Themes of semi-directed interviews with interviewees*

3) After interview

It appeared important to maintain contact with the patients, at this point, for two main reasons: to value their experiences as part of a sharing of expanded knowledge and a co-construction of knowledge (patient–researcher); a report on ethical and epistemological analysis could be conducted from this experience and knowledge.

In addition, the trust relationship established with patients, patient associations and physicians involved in the study enabled fruitful exchanges of experience during and after the interviews, in the case of telecare of cardiac and dialysis patients, in particular, establishing some ideas for future collaboration.

DEFINITION 6.1.– The role of a patient association is to break the isolation that disease can cause and make known or recognize often unfamiliar disorders. But gradually, patient associations became a partner for doctors and health institutions. They ensure good medical information to the patient, collect funds for medical research, participate in clinical studies and provide opinions on the care offered by doctors[14].

## 6.2. Wearing a heart defibrillator

### 6.2.1. *"Les Porteurs" patient association*

"Les Porteurs" patient association, chaired by Anne Pinna, aims for the development of a "chain of solidarity" between present and future members, as well as their families, presenting high-risk disorders and carriers (or previous ones) of a cardiac stimulator (pacemaker) or a defibrillator. Sharing experiences and testimonies, member information on new technologies, medical news, mutual assistance, collective or individual defense for its members, medical and social information requests alongside professionals and participation in prevention are the drivers of this association[15].

### 6.2.2. *Association de Porteurs de défibrillateurs cardiaques*

The *Association de Porteurs de défibrillateurs cardiaques* (APODEC, Cardiac Defibrillator Carrier Association), founded in 2000 by a small group of defibrillator

14 Source: allo docteurs. https://www.allodocteurs.fr/actualite-sante-quel-est-le-role-des-associations-de-patients-_1703.html.

15 Source: http://lesporteurs.fr/but-de-lassociation-les-porteurs/. Page consulted on April 1, 2016.

carriers and chaired by Alain Perrin since 2016, strives to be a "chain of solidarity between present and future patients with high-risk rhythm disorders". Sharing experiences and testimonies, mutual assistance, collective or individual defense, medical and social information requests alongside professionals and participation in prevention and fighting against all means of sudden death are the drivers that constitute this association. One of APODEC's struggles was allowing anyone presenting a risk of sudden death to aspire to have a defibrillator implanted and thus protect them from certain death. For this, it was necessary that social security cover the defibrillator.

Today, the APODEC works on the systematic management of telecardiology for each implantee because it is aware of the benefits of telecardiology for carriers of defibrillators, namely, *from the medical point of view*: a reduction by half of mortality, a reduction of almost 75% of the number of hospitalizations, a 52% reduction in appropriate shocks; *from a quality of life point of view*: a positive psychological aspect, telemonitoring facilitates the acceptance of implantation, reassures carriers and families, means a freer life compared to the disease, fewer monitoring visits that are often useless when all is well, less travel which is sometimes very long, less days off to ask from employers, less fatigue[16].

### 6.2.3. *Expert patients*

DEFINITION 6.2.– An expert patient means one who has a chronic condition and has developed over time a fine knowledge of their illness and has learned to live with it. The expert patient is above all an actor of their own health, but can also act as a resource for others[17].

The so-called "expert" patients are generally present in patient associations, participate in seminars and various media, perform an intermediary role between patients and physicians as part of meetings organized in the hospital workplace and may occur as part of mixed teachings (patients/doctors) provided in patient universities.[18] Expert patients have valuable experience of disease and telemonitoring that allows them to be at the crossroads of the needs and expectations of chronic patients and healthcare professionals.

16 Source: http://www.catelparis.fr/remuneration.html. Page consulted on April 1, 2016.

17 Source: health project. http://projetsante.com/Le-patient-expert/.

18 The University of Patients is an innovative educational set-up that consists of incorporating academic degree courses in therapeutic education of expert patients from the associative world. Source: http://www.universitedespatients.org.

### 6.2.4. *Pacemakers and implantable automatic defibrillators*

**Pacemakers**

A pacemaker is implanted in the event that the heart is too slow (bradycardia), generally less than 50 beats per minute for a non-athlete. It will thus stimulate the heart, if necessary, so that it is kept at a normal heart rhythm. Of course, this heart device has an electric battery. As for implantable defibrillators, the doctor will adapt different settings (thresholds) to the device according to the pathologies of each patient. Note that pacemakers are smaller than defibrillators because they do not need to store energy to deliver shocks (in capacitors). They weigh about 25 g and are 6–8 mm thick.

**Box 6.2.** *Description of pacemakers*

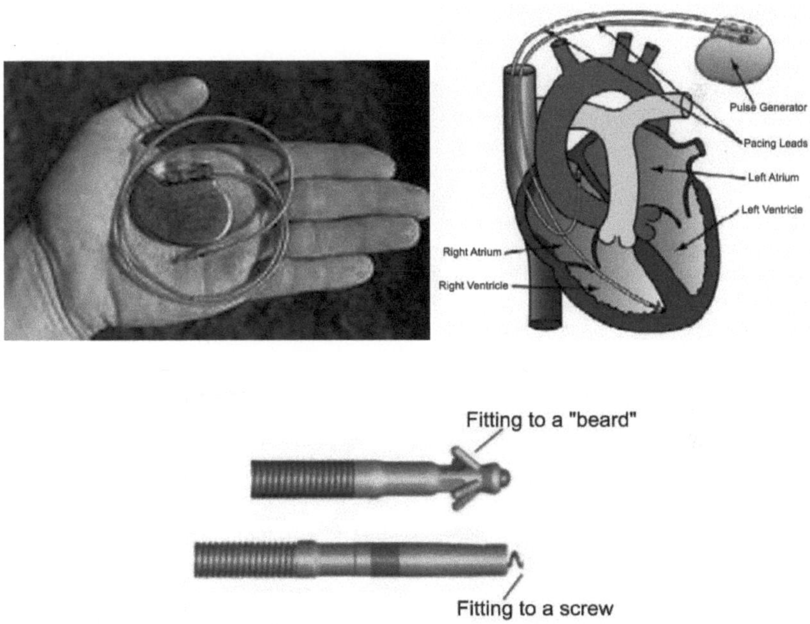


**The different types of fittings**

**Figure 6.1.** *Pacemakers*[19]

19 Source: http://lesporteurs.fr/les-stimulateurs-cardiaques-pacemakers/. Page consulted on June 10, 2016.

**Implantable automatic defibrillators (IADs)**

IADs are a little larger than a pacemaker: 50–80 g with a thickness of 9–15 mm.

The IAD is intended to correct tachycardia (rapid acceleration of the heart) in two different ways: either by stimulating the heart to stop tachycardia, or by delivering an electric shock if tachycardia does not stop or exceeds certain thresholds set by the rhythmologist (number of beats per minute, tachycardia, etc.), or if the heart begins to beat in a completely haphazard and inefficient way (ventricular fibrillation) because this irremediably leads to cardiac arrest and death if no shock is administered within 3 min. It is up to the rhythmologist doctor to adjust the settings of the various thresholds depending on the pathologies of each patient. Note that all the defibrillators also serve as a pacemaker because they stimulate the heart in case of bradycardia (slow heart).

**Box 6.3.** *Implantable automatic defibrillators (IAD)*[20]

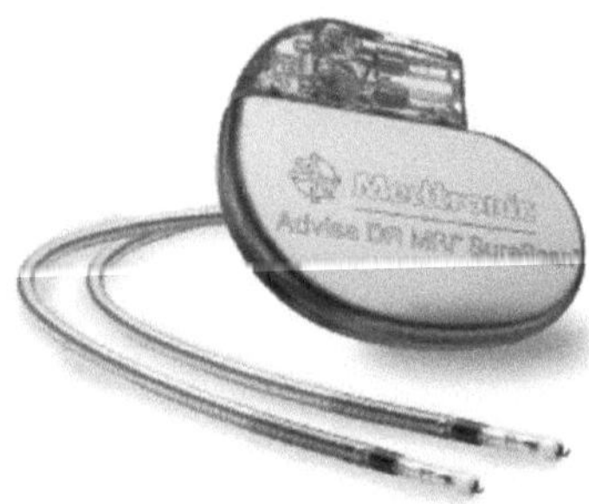

**Figure 6.2.** *Model of a Medtronic defibrillator for heart failure (implanted)*[21]

**Technical specifications of pacemakers and defibrillators**

Pacemakers and defibrillator boxes are connected to a (single chamber) or two probes (double chamber) located in the atrium or right ventricle via veins (usually passing under the collarbone). The probes are fixed in the heart by a screw system with a "beard" system (like an anchor, see Figure 6.1). We can also now find three probe (triple chamber) devices to treat heart failure via a resynchronization of the left ventricle (which sends blood in the arteries). However, this third probe is fixed outside the left ventricle (probes cannot be implanted in the atrium or left ventricle because blood flow is too strong and there is a risk of a blood clot).

20 Source: http://lesporteurs.fr/les-defibrillateurs-automatiques-implantables/. Page consulted on June 10, 2016.

21 ADVISA A3DR01 Model with Medtronic SureScan RM probes. Courtesy of Medtronic. Other brands in this market are Biotronik and St. Jude.

The life expectancy of this heart device is about 7 years on average. Of course, every time it goes off (stimulation) that lowers life expectancy. When the minimum level of battery charge is reached (around 4%), the defibrillator must be changed: the patient must then undergo a new implantation. The operation takes about 1 h.

The rhythmologist must regularly check (every 3 or 6 months) the functioning of the boxes to make sure that the probes are in good working condition and are unlikely to cause unexpected stimulation (in case of breakage or wear), that the battery is in good condition and is not at the end of its life, or to check the events recorded by the device's memory (rhythm disorders, for example). These devices act as a "holter" because they record cardiac events that have occurred since the last check (depending on the settings fixed by the doctor). The holter is only responsible for the detection or monitoring of heart rhythm disorders. It is indicated whenever the doctor thinks that a heart rhythm abnormality is responsible for symptoms or to assess the effectiveness of a treatment or the functioning of a heart pacemaker. It is therefore mainly used for diagnostic purposes.

*Precautions to be taken after implantation*: it is essential to avoid going near any device emitting electric waves such as induction plates, airport metal detectors, electric arcs (such as arc welding), etc. It is advised to present a carrier card issued by the implant doctor. For store alarms, it is recommended not to stand between them. For sport precautions, seek a cardiologist who gives recommendations based on pathology. Of course, all violent sports must be avoided (rugby, judo, etc.) or sports that are very demanding on the left arm (in the case of box implantation on the left-hand side) to avoid the risk of premature wear or the breaking of sensors. If a patient needs a Magnetic Resonance Imaging (MRI), there are precautions to take based on the box models (MRI compatible or not). It is essential to speak with a cardiologist and radiologist.

**Box 6.4.** *Technical specifications of pacemakers and defibrillators*[22]

### 6.2.5. *Standard protocol*

These computerized and connected prostheses (according to models) are subject to a one-time telemonitoring term calendar and/or continuous special event management. The functionality of these devices is simple: continuous stimulation of the heart in the case of the pacemaker, stimulation and defibrillation ("appropriate shock")[23] in the case of arrhythmia or cardiac arrest in the case of the defibrillator.

22 Source: http://lesporteurs.fr/les-stimulateurs-cardiaques-pacemakers/. Page consulted on June 10, 2016.

23 A shock is "appropriate" when it is caused by the defibrillator in reaction to a proven cardiac event. We talk about "inappropriate" shock when the shock is triggered by an error from the defibrillator.

The patient cardiac data transfer modes recorded by the pacemaker or defibrillator are of two types: either by teletransmission (Biotronik defibrillator, for example), every 24 hours, by saving the data in the defibrillator memory card (St. Jude model, for example), with, in this case, an extraction and a regular reading by the rhythmologist–cardiologist. Teletransmission involves the installation of a transmission box connected to the patient's home. This box is not required in the case of an IAD equipped with a memory card. The care protocol is standardized and, except for complication, is as follows: follow-up visit every 3–6 months, on average, at the town cardiologist's office, and consultation with the rhythmologist–cardiologist one to two times a year, on average. Personalized daily drug treatment of an anxiolitic and beta blocker type is added to this monitoring. In case of complications, the cardiologist is informed in real time, and depending on the severity of the alert (yellow, orange or red), can contact directly, via SMS or by phone, the patient to whom they are responsible.

Various software applications and communicating objects can also be used by the patient, in addition to the pacemaker or defibrillator as part of their telemonitoring, to get more information on their state of health: tablet, PC, smartphone, smartwatch, companion software.[24] The patient may use these objects and health applications to track vital signs themselves and some physical parameters (number of steps taken daily, caloric intake, followed by sleep intake).

### 6.2.6. *Experience of the protocol*

From the interviews, it can be noted that telemonitoring is experienced differently by each carrier according to the age, the type of implanted defibrillator, shocks already suffered and the quality and frequency of appointments with the rhythmologist–cardiologist. Overall, the town cardiologist is the trusted person in terms of proximity who can meet any challenges faced by the carrier. The rhythmologist is a more remote professional who analyzes the data and gives meaning to the cardiac episodes having been experienced by the patient. Nurses provide a relay role in the sense where they collect daily data from the defibrillator and report problems to the rhythmologist who intervenes, as appropriate, at the earliest time possible. In certain hospitals and cardiology services, psychologists are put at the disposal of patients who feel the need to talk about their physical health, their distress and who wish to be advised and accompanied[25]. Overall, psychological assistance proposed today is, however, not enough in the opinion of carriers.

---

24 The ap'Telecare type (see Chapter 5).
25 This is the case, among other things, of the cardiology service from the University Hospital of Lyon.

Nursing home visits are rare, and are not really justified in themselves. Family caregivers play a central role in terms of patient support. They can provide major emotional and logistical support. The anxiety of caregivers and loved ones can however be a burden, according to certain cases. We therefore speak of the *transfer of concern*: the restlessness of caregivers in addition to that of the patient. The origin and the degree of anxiety for caregivers particularly depends on woven emotional ties with the patient. Some carriers feel rather anxious and stressed, because of their pathology, and wish to have more information on their state of health. Some people suffer, for example, during the first implementation, from a lack of information. They wish, in particular, to be better informed about the nature of "shocks" that can be triggered by the defibrillator and how to manage them at best. Some carriers, on the contrary, prefer to maintain their distance with telemonitoring by seeking not to be too informed, and are eager to optimize the terms of autonomy provided by the pacemaker-defibrillator. In principle, carriers can follow therapeutic rehabilitation sessions (one to two sessions per year)[26]. Many cardiologists, in the general opinion of defibrillator carriers, are insufficiently aware of therapeutic education.

Two major sources of stress are also identified by all carriers: on the one hand, the possibility of the defibrillator probe breaking and, on the other hand, the possibility of a heart transplant. The probe(s) of the defibrillator is(are) in effect transplanted onto the flesh (arteries), or even the heart. Changing a probe is an extremely delicate operation, lasting several hours, requiring general anesthesia. Moreover, the heart transplant is not trivial. The operation has risks, particularly rejection. In addition, the transplant is done as an emergency. Patients effectively have a very short time (about 4 h) to reach the hospital for the procedure.

Regardless of their activity or their geographical location, all surveyed carriers recognize a great personal benefit from their membership in a patient association: it promotes the exchange of experience in a friendly atmosphere and allows, if necessary, the patient to break isolation.

Eleven defibrillator carriers and members of the two previously noted patient associations agreed to the proposed interview: seven men and four women with an average age of 57 years. Table 6.3 lists the main biographical data (anonymized) of the interviewed defibrillator carriers.

---

26 Therapeutic education, outside the hospital, is provided by physiotherapists, nurses and rhythmologists. It particularly consists of a stress test and sports activities (cycling, treadmill, physical education).

| Patients | Age | Activity | Pathology and date of diagnosis | Implanted defibrillator model | Implant center/date of implantation | Defibrillators implanted since diagnosis |
|---|---|---|---|---|---|---|
| Mrs. D | 37 | Part-time professional | Arrhythmogenic right ventricular dysplasia (1994) | Saint-Jude | St Joseph Clinic – Marseille (13) 2009 | 1 |
| Mr. T | 47 | Full-time professional | Cardiac arrhythmia (2008) | Biotronik | Cavale Blanche Hospital – Brest (29) – last implantation 2008 | 2 |
| Mr. L | 70 | Retired | Cardiac arrhythmia (2008) | Medtronic | Pinel Hospital – Amiens (80) 2008 | 1 |
| Mrs. C | 50 | Full-time professional | Cardiac arrhythmia (2016) | St. Jude | Avignon Hospital (84) 2016 | 1 |
| Mr. G | 63 | Full-time professional | Brugada syndrome (2000) | Biotronik | Pasteur Clinic – Toulouse (31) – first implantation: 2000 | 4 |
| Mr. G | 52 | Disabled | Genetic insufficiency | Medtronic | Lyon Hospital (69) – last implantation: 2008 | 2 |
| Mr. B | 58 | Full-time professional | Cardiomyopathy (2009) | Biotronik | Bordeaux Hospital (33) 2011 | 1 |
| Mr. P | 70 | Retired | Genetic insufficiency (2001) | Biotronik | Croix-Rousse Hospital (69) – last implantation 2013 | 3 |

| | | | | | | |
|---|---|---|---|---|---|---|
| Mrs. P | 54 | Full-time professional | Genetic insufficiency (2008) | Guidant Vitality (Boston Scientific) | Rieux Hospital (56) 2008 | 1 |
| Mr. G | 65 | Retired | Heart attack (1996) | Sorin | Bichat Hospital (75) – last implantation: 2011 | 2 |
| Mrs. J | 60 | Part-time professional | 2002 | Guidant Vitality (Boston Scientific) | Mercy/Metz Hospital (57) – last implantation: 2009 | 2 |

**Table 6.3.** *Biographic data of interviewed carriers*

### 6.2.7. *Carriers' words*

The tables proposed in the following pages reflect the feelings of the defibrillator carriers. These tables represent an (assumed) alternative to the full transcript of interviews conducted with carriers. They make it possible to give priority to the word of the patients, according to the central issue of the book, avoiding two major pitfalls: relegation to the annex, on the one hand, of a full transcript of the interviews and the offer, on the other hand, of a synthesis that would fail to take account of people's "plain speech".

The selection criteria of opinions and comments of the interviewed carriers are based on their recurrence factor: the objective of the study is to identify the main trends in the feeling of telecare expressed by the patients. *Giving patients a voice* presents here two advantages: on the one hand, this approach has a "didactic" function by inviting the reader to share *the reality of disease and telecare as expressed by patients*. Without reaching the degree of proximity that the "live voice" interview allows, this approach, nonetheless, enables us to identify people's feelings; on the other hand, it sheds light on the synthesis of this work to be proposed later.

Three impact types of telemonitoring on cardiac patients were mainly assessed: physical, temporal and psycho-emotional. Three complementary aspects of telemonitoring, providing a broad perspective of the feeling of interviewees, have also been taken into account in the analysis, namely:

1) the specific effects of communicating objects;

2) the aspects of telecare to improve according to the patients;

3) the *representative* reviews and comments of the patients, namely the recurring ones, within the studied group.

| I) Positive *physical* impacts |
|---|
| 1) "The defibrillator is completely part of me. It isn't a foreign body. I've tamed it" (Mrs. J). |
| 2) According to the pathology, and the implanted device, the defibrillator can be "forgotten" over time (Mr. T). |
| 3) "My defibrillator has saved my life" (Mrs. D). |

| I) Negative *physical* impacts |
|---|
| 1) It implies an athletic activity or training rhythm change. "The defibrillator has changed my relationship in terms of effort required. I should avoid being short of breath, I should not force it. I need to go gently" (Mr. P). |
| 2) The defibrillator can trigger "inappropriate shocks" (anxiety-inducing) (Mr. T). |
| 3) You must constantly watch out for the defibrillator: avoid shop alarms or airport metal detectors, sports that are too violent, avoid "heavy" do-it-yourself work, avoid raising the arms too much, etc. (Mrs. C). |
| 4) The defibrillator can move within its box (which is very stressful) (Mrs. D). |

| II) Positive *temporal* impacts |
|---|
| 1) High reactivity, or even proximity, of the cardiologist in case of problems: "I can call him anytime, I have his home phone number" (Mrs. D). |
| 2) No change in daily activities since the positioning of the defibrillator, even if the pace of life can sometimes be slower. |
| 3) Certain problems can be detected through telemonitoring before they affect health status. |
| *Before telecardiology, information was stored in the pacemaker/defibrillator and read every 6 months, at each appointment with the specialist. With telecardiology, it is now possible to save weeks, or even months, in the detection of an anomaly. This can prevent serious complications among patients. In addition, they do not have time to worry since they haven't had the time to feel any symptom.* |

4) Lifetime of the defibrillator (7 years on average, according to the model).

5) Reduction in the number of visits to the cardiologist and hospital (only 2 visits per year on average).

6) The data collected daily from the defibrillator, reported by the rhythmologist, are reassuring and useful.

*The information provided in the report includes tachycardias, heart rhythm disorders, probe impedance and the general status of the device. The copy of the statement is sent to the treating physician. But the set of all statistical data is not provided to patients. The recorded information is reassuring and sufficient as it allows us to answer the main questions that can be asked by the carriers: what were the events recorded by my defibrillator? Is it necessary to change the treatment, revisit the doctor, or do blood tests?*

7) The data collected remotely on the defibrillator allows us to regularly check the charge level of the prosthesis, the status of the probe and/or the device.

II) Negative *temporal* impacts

1) The lack of availability from hospital staff to perform the daily summary of data transmitted by the defibrillator, which can result in a delay in the follow-up to the alerts.

2) The frequency of contact with the rhythmologist–cardiologist may be insufficient at times.

III) Positive *psycho-emotional* impacts

1) *The presence of the defibrillator is lighter over time.*

2) The ability to maintain a professional activity and therefore "forget" the constraints of the disease.

3) "The sense of security linked to the fact of being telemonitored 24/7" (Mr. B).

*Today, the defibrillator saves and sends all the information, on a daily basis, without requiring the patient to intervene. It is a matter of checking that the patient is well, but also that the prosthesis is in perfect working condition. These data are added on a daily basis, automatically, to the secure patient record. If an anomaly is detected, the cardiologist who follows this person will receive an alert via SMS, for example, and make contact, as appropriate, with the patient.*

4) The peace of mind provided by the defibrillator: the disease can be, depending on the case, kept at a distance.

5) Therapeutic rehabilitation enables better management of one's daily routine as a carrier (Mr. G).

| |
|---|
| 6) Discussion with the rhythmologist during semi-annual or annual review sessions. |
| 7) The defibrillator provides protection against sudden death: "The defibrillator is my guardian angel" (Mr. G). |
| 8) "I have a device, everything is alright. Why be stressed? On the contrary, the device is there to save me" (Mr. B). |
| 9) "I am eminently aware of how lucky I am to be alive. It is something that lives with me since my episode of sudden death" (Mrs. J). |

| III) Negative *psycho-emotional* impacts |
|---|
| 1) The intrusiveness of the defibrillator (at least during the first few years). Difficulty in accepting the implanted device (Mrs. J). |
| 2) Lack of information and psychological support after implantation (Mrs. J). |
| 3) Lack of psychological support by a practitioner trained in cardiac pathologies (Mr. G). |
| 4) Difficulty recovering from a shock triggered by the defibrillator when you get older (Mr. G). |
| 5) Uncertainty about the risk of a cardiac incident, which is one of the main factors of stress. |
| 6) Lack of awareness about therapeutic education among some city cardiologists (which can be stressful for the patient). |
| 7) Change in "character" after implantation (increased susceptibility, enhanced bad mood). |
| 8) The defibrillator manufacturer's instructions are alarming and may result in denial of the device. "When you read the manufacturer's manual, it's scary" (Mr. L). |
| 9) Shocks (appropriate and inappropriate) are a constant threat. |
| 10) The fear of an operation to change a broken sensor, because it is a delicate operation (Mr. G). |
| 11) "I have something to reproach the cardiologists about: they are not listening to patients" (Mrs. C). |
| 12) The experiences of the disease and the defibrillator are better understood by the patients and doctors who have experience and knowledge: "Relatives can not really understand" (Mr. G). |

13) Support from loved ones can cause *a transfer of concern.*

14) According to lifestyle and/or professional activity, it may be difficult to accept/manage the pacemaker or defibrillator.

15) Putting oneself in professional overactivity to forget the stressful effects of the defibrillator and pathology (Mr. G).

16) Wearing a defibrillator has many drawbacks: the device can be seen and is not aesthetic (problem from the perspective of the other person), it can hurt and have an impact on sex life.

17) "What concerns me is what this *[wearing a defibrillator]* does to the spouse" (Mr. B).

18) "Counseling helps us morally and psychologically. But it cannot be imposed" (Mrs. J).

19) Some tests done at the hospital, required by the health insurance fund, may be stressful (Mr. G).

20) "The rhythmologist does not necessarily call to make the point in case of 'small' recorded discharge from the defibrillator. This surprised me at first because I was expecting him to do it routinely after each shock" (Mr. P).

21) It can be difficult, in the early days, to accept the defibrillator (it is an intruder).

IV) *Positive* impacts of the communicating objects

1) The use of a connected home scale to track daily weight, associated with a therapeutic education, is reassuring.

*The Cardiauvergne association proposes telemonitoring of weight through the installation of a connected home scale. On a larger scale, Cordiva is a home surveillance program for heart failure. It is at the center of two clinical reference studies currently being conducted: Osicat (by the Toulouse University Hospital) and PIMPS (by the Pontoise Hospital). More than 16,000 patients are monitored by Cordiva in Germany, and the first results have already confirmed positive effects on mortality and on the medical aspects of the disease. The Gesellschaft für Patientenhilfe study (AOK data) in particular noted a 28.3% reduction in hospitalizations in patients following the program. Cordiva is a telemonitoring device that provides an electronic scale and a connecting box by post to patients' homes, accompanied by documentation on heart failure. Patients should weigh themselves daily and answer a few questions posed by the communicating box. The data are then automatically transmitted to the tele-support center. The center is controlled by specifically trained nurses and in contact with healthcare professionals caring for the patient. The center is the focal point of information concerning the patient. Cordiva monitoring combines clinical data transmitted by the connecting box and an education in terms of the patient's health. This health education is personalized based on the degree of patient knowledge and the evolution of their heart failure. Every day the clinical data of the patients (weight and answers to the questions) are analyzed by a computer system that triggers alerts according to defined criteria*[27].

2) Use of a smartwatch and/or a smarpthone equipped with a health application, of the "PACER" type[28], "helps reassure and measure the effort" (Mr. G).

3) The absence of a transmission box (in this case, the "communicating" defibrillator is equipped with a built-in memory card). "It allows us to not think about the illness on a daily basis" (Mrs. D).

4) The alert system warns the doctor in real time.

| IV) *Negative* impacts of the communicating objects |
|---|
| 1) The use of the *defibrillator as a communicating* object can contribute to the impersonalization of care: "Telemonitoring does not necessarily manage the emergency: in fact, it manages the battery [*of the defibrillator*]. It alerts the patient of something that they have already felt themselves" (Mr. B). |
| 2) Human dimension is sometimes lacking: "Telemedicine indicates less human contact therefore it must be compensated by something else" (Mr. T). |
| 3) Lack of training in prosthetics of certain practitioners. Interns do not necessarily control all defibrillator settings and can make damaging mistakes for the patient (Mr. G). |
| 4) The disadvantages of the warning system. It is not known if the data transmitted to the care center by the defibrillator connected to the home terminal are verified/analyzed daily (Mr. G). |
| 5) "Connecting every day with the transmission box is a constraint. Some patients are choosing to disconnect themselves" (Mrs. P). |
| 6) "I had a home transmission box but I always went to the hospital because I had constant interference with my landline" (Mrs. J). |

27 Source: Sébastien Woynar, "Surveillance à domicile du patient insuffisant cardiaque : un accompagnement sur la durée et un confort au quotidien": presentation at the summer University of E-health, July 5–7, 2016 in Castres. Source: http://www.universite-esante.com/session/alere-cordiva/. Page consulted on July 5, 2016.

28 A management software for health on iPhone, combining the pedometer, weight management and recording of blood pressure.

| V) Areas for improvement in cardiac telemonitoring | |
|---|---|
| Psychological support | 1) Bridging the information gap after implementation. "I got out of my second implantation, and they didn't even tell me what they had put in [defibrillator model]" (Mr. G).<br><br>2) Be better trained in the use and the workings of the defibrillator and receive psychological support after implantation and/or during the adaptation phase (Mr. G).<br><br>3) Benefit from counseling provided by a professional trained in cardiac pathologies: "I have more need for the human eye, because I already have the medical eye" (Mr. T).<br><br>4) Offer psychological support to family caregivers (Mr. G).<br><br>5) Create thought-provoking questions on the relational dimension of telecare (Mr. B).<br><br>6) Encourage exchanges of experience between patients (awareness of patient associations) and between patients and healthcare professionals (Mrs. C).<br><br>7) Involve the carriers and/or expert patients in rehabilitation sessions, even after the first implantation to inform and reassure (Mr. T). |
| Telecare technologies | 1) Ensure that broadcast data are analyzed daily by a competent person (Mr. G).<br><br>2) Be able to collect a daily return from the rhythmologist on the data collected (Mr. G).<br><br>3) A line open 24/7 to ensure medical and psychological monitoring of the patient: "We should have a line open 24/7 with someone competent to report why and how" (Mr. G).<br><br>4) The defibrillator is an "opportunity". "You must stop thinking about the pathology and tell yourself that there is worse than us" (Mrs. D). |
| Therapeutic education | 1) Therapeutic education must be developed: "If you want to delay the transplant, rehabilitation is the only way" (Mr. G). |

| VI) Attitudes and opinions of the carriers |
|---|
| 1) Take a step back. Practice meditation and "mild" sports: Tai Chi, for example. "Stress is very difficult to manage: it's one of the challenges for medicine in the future" (Mr. T). |
| 2) Accept the slow pace: "I put my foot on the brake and accept it" (Mr. T). |
| 3) Cardiologists should be encouraged to build a network of partners so that they can provide an appropriate response to questions from carriers (Mr. T). |
| 4) "When you have a serious pathology, you must distance yourself. But some people do not always do that" (Mr. B). |
| 5) "The danger of telecardiology is to cut a true, authentic connection, in reality, with the team. Boxes and smartphone applications will not replace human contact. It is an illusion to think that" (Mrs. J). |
| 6) You should not be able to cancel the half-yearly appointment with the cardiologist because this is the only time where we have a little info (Mr. G). |
| 7) Cardiologists must speak clearly to their patients, explain the pros and cons of the defibrillator and home care; they mustn't leave the disadvantages aside. The patient should not have to learn these disadvantages following the facts (Mr. P). |
| 8) "It's important not to judge people in the way they react to the disease" (Mr. B). |
| 9) "Special situations (social, family and professional) of carriers to better answer their questions must be considered" (Mr. G). |
| 10) "After implantation, a life project must be given to oneself. Rebuild oneself, then be a witness to others" (Mr. G). |
| 11) "I think deeply that one must make friends with his defibrillator" (Mrs. J). |
| 12) "We're completely in tune with death. We're in a matter of life and death" (Mrs. J). |

## 6.3. Performing daily home dialysis

### 6.3.1. *"Renaloo" patient association*

Renaloo was established in 2002, in the form of a blog, becoming a patient association in 2008. The association is developing many activities, on and off the internet, to more effectively convey its support and empowerment values for people living with kidney disease, dialysis, transplantation, defending their rights and their interests and improving their care and their lives. The association's mission is based on the commitment of the Renaloo team, consisting of people living with kidney disease and loved ones who have chosen to get involved[29].

### 6.3.2. *"Calydial" association*

As an Établissement de Santé Privé d'Intérêt Collectif (ESPIC)[30] – a not-for-profit association under the French law of 1901 – created in 1984, Calydial is governed by the FEHAP collective agreement of October 31, 1951. Calydial's care activities are authorized by the Rhône-Alpes ARS (regional health agency). Calydial is a health care facility specializing in the treatment of patients with chronic kidney disease. Calydial's goal is to: "Make possible all forms of substitution treatments for patients with chronic renal failure, provide patients with more self-sufficient means of treatment, adapted to their state of health, gather interested people and those involved in all problems posed by this type of treatment"[31]. Doctor Agnès Caillette-Beaudoin in particular, nephrologist and medical director at the Calydial dialysis unit established in Irigny (69), has given her support to this research and allowed telemonitored interviews of patients in independent home peritoneal dialysis.

### 6.3.3. *Tenon Hospital (AP-HP- 75) and other actors*

This establishment is the first French care center and the second European Center in terms of number of patients followed in independent daily home hemodialysis[32]. Doctor Hafedh Fessi, nephrologist at the Tenon Hospital (AP-HP) interviewed for the purpose of this book, is, with Doctor Agnès Caillette-Beaudoin mentioned previously, one of the main instigators of the development of telemonitoring of independent daily home dialysis in France today.

Many testimonies have been collected on the internet, especially on the patient association websites (Renaloo), some thematic channels (YouTube), telemedicine industry sites (Theradial.com), the *Ligue Reins et Santé [Kidneys and Health League]* association and various blogs of independent home dialysis patients[33].

29 http://www.renaloo.com/renaloo/renaloo-en-association. Page consulted on July 10, 2015.

30 A private health establishment of collective interest.

31 See also the description of Calydial in Chapter 4.

32 http://tenon.aphp.fr/nephrologie-dialyse/.

33 See the websites and blogs of patient experts listed in the webography.

### 6.3.4. *The cyclers and communicating objects*

**Figure 6.3.** *Cycler for independent home dialysis*[34]

The principle use of the cycler is simple. This is a compact but heavy (35 kg) device allowing positioning of a kind of tape that contains the lines, the dialyzer, and hanging dialysate pockets already ready to use. The preparation of the machine takes less than 15 minutes to measure blood pressure, weigh yourself, and register data in the dialysis journal (or electronic dialysis notebook)[35].

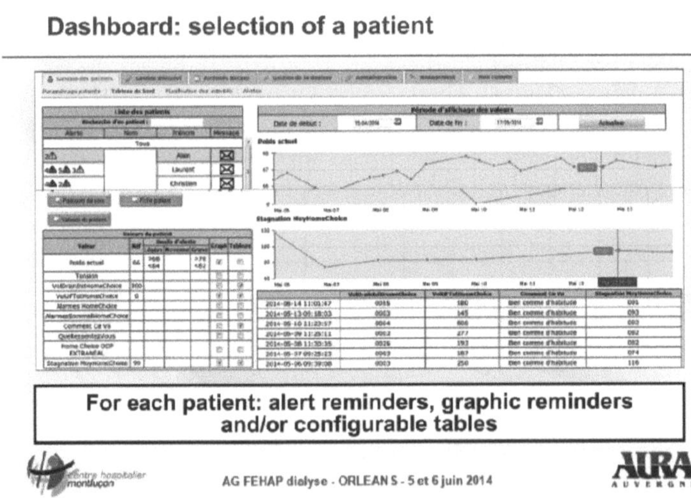


**Figure 6.4.** *Screenshot of the health application "ap'Telecare"*[36]

34 NxStage System One model of the NxSTAGE company. Courtesy of Nx STAGE. The other large brands on the cycler market are Nikkiso, Tauro Pharm and Serumwerk.
35 [REI 14, p. 27]. For a description of the dialysis procedure, see section 4.2.
36 Source: [DAB 14].

### 6.3.5. *Standard protocol*

The technique used for home hemodialysis is similar to all other hemodialysis procedures. Only monitors differ somewhat, the machine being easier to handle, with additional safety features to avoid handling errors. Indeed, this procedure completely ignores any healthcare professional, the session being fully managed by the patient themselves, or by a member of their entourage. Usually, connection is done by the medical-venous fistula bipuncture in a "button hole" technique that the patient must learn to do themselves, or have a member of their entourage carry out[37]. During a puncture accident, dialysis can also be done in unipuncture, with a "ClicClac" rescue kit.

More rarely, vascular accessibility is done on a central catheter in double light, with strict observance of all rules of hygiene and sterility. The implementation of home hemodialysis requires rigorous teaching with an obligatory stay in the self-care unit. The patient must be able-bodied, willing to request help and fully aware of the constraints and risks that the procedure involves. Their home must also be adapted. Training usually takes two months. During this stage, all the technical adaptations are performed at home (choice of a room with suitable water supply and drainage), and the attending physician is consulted and informed.

**Box 6.5.** *Description of a standard protocol for home dialysis [REI 14, p. 27]*

Medical telemonitoring allows the coordination of care for home dialysis. Slowing the progression of chronic renal failure is achieved through controlling factors such as blood pressure and diet for which the use of communication tools is particularly relevant. The possibility of monitoring these indicators mainly avoids accelerations of the chronic disease[38]. In addition to the cycler, some independent home dialysis patients, including those interviewed for the purposes of this book, may use an electronic dialysis diary: because of their e-tablet, patients can validate the stages of the course of dialysis and relay all information to the healthcare team (exchange of peritoneal dialysis, diabetes monitoring, validation of the injection of the erythropoiesis stimulating agent)[39].

---

37 "With *the button hole*, there are two points which are always in the same place, however, a high level of hygiene is required and the passage and the valve in the tunneled part must be found. There is no sharp and pointed needle to tear the fistula, they are instead needles with foam tips, and if one is not in the passage, they do not fit into the fistula, do not cause a hemetoma or large aneurysm (...). The advantage of self puncturing and *the button hole* is that there is no pain" [REI 14, pp. 29–30].

38 See the description of the Calydial home dialysis protocol in section 4.2.

39 For a detailed presentation of the electronic dialysis diary running ap'Telecare, see also section 5.3.2. For a definition of erythropoiesis, see p. 119.

### 6.3.6. *Interviewees*

Five patients with chronic terminal renal failure in independent home hemodialysis, telemonitored by Doctor Fessi and his team at the Tenon Hospital (75), as well as four patients on peritoneal dialysis followed by the Calydial association, accepted the proposed interview: seven men and two women with an average age of 56. Table 6.4 lists the main biographical data (anonymized) of these people.

| **Patients** | **Age** | **Activity** | **Date of diagnosis** | **Hospital – auto-dialysis start** | **Cycler model – dialysis system** | **Mode of telemonitoring** |
|---|---|---|---|---|---|---|
| Mr. S | 40 | Part-time professional | 2014 | Tenon Hospital (75) 2015 | PHYSIDIA | Physidia tablet and software |
| Mrs. D | 39 | No professional activity | 1998 | Tenon Hospital (75) (2016) | PHYSIDIA | Physidia tablet and software |
| Mr. F | 43 | Full-time professional | 1990 | Tenon Hospital (75) 2014 | PHYSIDIA | Tablet and software (Physidia) |
| Mr. V | 67 | Retired | 2000 | Tenon Hospital (75) 2014 | PHYSIDIA | Tablet and software (Physidia) |
| Mr. F | 71 | Retired | Congenital pathology | Kidney Clinic (69) 2015 | Infusion pole + dialysate pockets | apTelecare (Calydial) |
| Mr. R | 59 | No professional activity | 2015 | Lyon Sud Hospital (69) 2014 | Infusion pole + dialysate pockets | apTelecare (Calydial) |
| Mr. M | 80 | Retired | 2013 | Portes du Sud Vénissieux Clinic (69) (2013) | Infusion pole + dialysate pockets | apTelecare (Calydial) |

| | | | | | | |
|---|---|---|---|---|---|---|
| Mrs. B | 47 | Disabled | 1982 | Tenon Hospital (75) 2015 | PHYSIDIA | Tablet and software (Physidia) |
| Mr. C | 58 | No professional activity | 2013 | Lyon Sud Hospital (69) 2015 | Infusion pole + dialysate pockets | apTelecare (Calydial) |

**Table 6.4.** *Telemonitored interviewed patients in independent daily home dialysis*

### 6.3.7. *Specifics of the protocol*

Independent daily home dialysis – a recent practice that concerns, to date, a small number of patients – differs from other existing modes of dialysis in France: either independent or assisted dialysis in dialysis centers and/or at the hospital. Treatment at dialysis centers involves frequent, even daily, travel for patients. In a dialysis center, these can be relatively passive sessions, without controlling their dialysis, that can be long (up to 4 or 5 hours a day). Home dialysis, however, allows patients to take care of their disease and their treatment, often at the times of their choice. Autonomous home dialysis helps patients in their own assessment of the disease and allows them to learn about themselves and to act on their own treatment. Independent dialysis, in which schedules are flexible (evening, morning or afternoon), can be an alternative to strong drug treatment. Several protocols can also be offered to all patients according to their activities and their health states: sessions of 4 hours, 3 times a week, or 2-hour sessions, six times a week, for hemodialysis (for patients monitored at Tenon Hospital). Rhythms of peritoneal home dialysis (monitored by Calydial) may also vary depending on the condition, age or activity of the person: between one and four sessions per day for 1 or 2 hours, on average and/or night sitting.

Another fundamental aspect of independent home dialysis (hemodialysis and peritoneal dialysis) is independent training provided by the care center, which can be several weeks. In this context, the nurse trains the patients in essential actions: pricking, use of dialysate pockets, the tablet, the cycler, etc. After this training stage, patients learn to manage their treatment and to independently manage the specifically technical part of care (cycler, pockets, input and sending of daily session summaries). The use of the communicating object as part of home hemodialysis (Physidia tablet model, for example) allows patients to remotely control some functions of the cycler (stop, standby, restitution, etc.) and transmit the information entered after each session to the medical team: voltage, weight, volume of the regurgitated pocket. The doctor and nurses receive real-time information and can view these data. At each monthly visit, these data are commented on by the nephrologist and medical examinations can be carried out to adjust, if necessary, the treatment according to the variation of patient data.

One of the main difficulties faced by dialysis patients, during the learning phase, is pricking oneself. One must be able to manage arteriovenous fistula. It is created by the surgeon who joins a vein in the arm in an artery, which has the effect of increasing the flow rate at this level and obtaining a large venous dilation and increased resistance of the wall of the vein, allowing point circuits for extracorporeal circulation during treatment with hemodialysis. The fistula has the advantage in terms of the catheter (placed on a large central vein in the case of peritoneal dialysis) of allowing healing between sessions and therefore less entry of germs.

Pricking can represent a real psychological barrier at the start. Even with experience, patients may feel some trepidation to accomplish this action. In fact, they always need to find the passage that has been created with the nurse for pricking. However, the fistula still moves a little bit: if the passage is moved a few millimeters, the action of the patient is different. Pricking is therefore a daily challenge. For some people, this can be an issue. In addition, as the action is "acquired", patients may have trouble letting the nurse do it (during monthly hospital consultations, for example): nurses indeed have an intimate knowledge of the fistula and can sense a certain apprehension that the action (pricking) is being carried out by them. The nurse is, nevertheless, trained to adapt to the patients; he/she can let the patient prick or guide and advise where to prick.

Independent dialysis patients are accompanied at all stages of their telemonitoring: upstream, for the preparation of the home dialysis area and storage of the equipment (consumables), during and after the installation phase and throughout protocol (medical, psychological, and technical support).

### 6.3.8. *Words from people in independent home dialysis*

The typology of the impacts of telemonitoring used for people in independent home dialysis is the same as that used for defibrillator carriers.

| I) Positive *physical* impacts |
|---|
| 1) "At the end of dialysis, when you're tired, you get up and push the machine button to return to [clean blood], while the tablet is next to me and I don't need to move, I can launch the return of blood while lying down" (Mrs. D). |
| 2) The possibility of pricking independently (Mr. M). |

| I) Negative *physical* impacts |
|---|
| 1) The change in sport activity or training pace (Mr. F). |
| 2) Nighttime peritoneal dialysis is not always possible because the machine is not as controllable as daytime dialysis (Mr. M). |
| 3) Dialysis four times per day limits travel. One can go on holiday, but only if one's vehicle is equipped (Mr. F). |
| 4) "There is a real psychological barrier when pricking. But after that, everything depends on one's level of motivation. There may be a concern later. One always needs to find the passage that has been created with the nurse for pricking. But the fistula still moves a little bit: it is sufficient that the passing moves a millimeter and the action is different. Pricking is therefore a daily challenge" (Mr. F). |

| II) Positive *temporal* impacts |
|---|
| 1) Certain problems can be detected through telemonitoring before they affect health status. |
| 2) The possibility to maintain or return to work and "set" home dialysis sessions (Mr. F). |
| 3) Responsiveness of the nurses, the doctor and the medical team before, during and after dialysis sessions (medical services provided until 10:00 PM) (Mrs. B). |
| 4) Reactivity of technical sustainability of the cycler manufacturer (Nx Stage or Physidia) (Mr. S). |
| 5) Time saving in terms of care (interfacing with the cycler and transmission of data from the session with the patient to the medical team) by using a simplified electronic tablet (Mrs. D). |
| 6) "Independent home dialysis allows you to save time, because there are fewer trips to the hospital or to a dialysis center" (Mr. S). |

| II) Negative *temporal* impacts |
|---|
| According to the experience and/or professional activity, it may be difficult to accept/manage the cycler (Mrs. D). |

| III) Positive *psycho-emotional* impacts |
|---|
| 1) My cycler is a part of the "family" (Mr. V). |
| 2) The wide availability of nephrologists and nurses is reassuring. |
| 3) Training workshops and the sharing of experiences implemented to teach the patient how to manage their independent home dialysis (Mr. V). |
| 4) "Before home dialysis, I took Lexomil, and the nephrologist made me meet a nurse psychologist and made me understand what was wrong. Since it has become much better and I no longer take anxiolytics" (Mrs. D). |
| 5) "Being on home dialysis, I feel less sick, it's not a hospital" (Mrs. D). |
| 6) Home dialysis allows the involvement of caregivers: "My husband felt powerless in the face of the disease. Suddenly, he participates and this also makes him feel good. He knows now what dialysis is. He feels useful and it's much better. This makes him feel good and me too" (Mrs. D). |
| 7) Reassuring fact: "To make a 'retreat' to the hospital in case of difficulties at home (for example, unable to find the passage when pricking)" (Mr. F). |
| 8) "With the psychologist [of the hospital], I can talk about the way I handled the alarms triggered by the cycler" (Mr. F). |
| 9) The hospital regularly offers a patient questionnaire on the feeling of independent home dialysis: this type of monitoring is comforting. |
| 10) The presence of the home care nurse (to replace the bandage from the catheter, for example) is reassuring. |
| 11) "The more you gain experience in the practice of care, the more stress decreases" (Mr. V). |

| III) Negative *psycho-emotional* impacts |
|---|
| 1) The fact that, initially, pricking by yourself at home can be stressful. |
| 2) "The alarm sounding in case of cycler malfunction generates stress: will I be able to do my dialysis? One must learn to manage the alarms: keep calm, keep cool. Nurses help us with this" (Mr. F). |

| IV) Positive impacts of the communicating objects |
|---|
| 1) The use of a connected scale at home to track daily weight, associated with a therapeutic education, can increase the feeling of security. |
| 2) Close remote monitoring of the cycler manufacturer (Nx Stage or Physidia) or hospital in case of technical problems with the cycler or physical problems encountered during dialysis. |
| 3) "With the dedicated tablet, I only manage my session" (Mr. F). |
| **IV) Negative impacts of the communicating objects** |
| 1) The cycler can be finicky and trigger alarms, in the event of malfunctions, during dialysis. |

| V) Areas for improvement in dialysis telemonitoring | |
|---|---|
| Psychological support | 1) Inform the patient at the time of diagnosis, all dialysis options and their impact on daily life. |
| Care equipement | 1) Waste collection service (used dialysis pockets) and removal of consumable cartons. |
| Facilities and healthcare professionals | 1) Each health care facility does not carry out diagnosis in the same way: there is a need to harmonize the practices. Good practices in this area must be identified: making a uniform announcement for all patients.<br>2) Information between nephrologists is not necessarily communicated. Some nephrologists sometimes withhold information (they do not inform patients of the existence of an independent home dialysis solution)[40].<br>3) Information on the patient between nephrologists and the implanters is not necessarily communicated. |

40 The lack of information found by some dialysis patients is not unique to France. A recent study conducted in Denmark on predialyses patients at home leads to a similar conclusion: "There is a real need for unbiased and structured information in predialysis, including sharing of patient experiences and the organization of treatment, that would allow home dialysis to be better exploited" [RYG 12].

| VI) Attitudes and opinions of people on independent dialysis |
| --- |
| 1) Take a step back and learn to manage stress. |
| 2) The nephrologists must speak clearly to their patients, explain the pros and cons of the various options of dialysis. |
| 3) There is a lack of awareness of independent home dialysis among some physicians and nephrologists. |
| 4) Excellent network of healthcare professionals involved in home care and monitoring of the person (doctors, nurses, nephrologists). |
| 5) "The transplant is a hope: it would be a big sense of relief" (Mr. M)[41]. |
| 6) "Independent dialysis is like drinking, eating, sleeping, as a routine activity. The dialysis must become part of the routine and the surrounding family must see this as normal" (Mr. V). |

## 6.4. Managing diabetes independently

### 6.4.1. *CERITD*

The non-profit association, CERITD (*Centre d'Études et de Recherches pour l'intensification du Traitment du Diabète*, French Center for Diabetes Treatment Intensification Studies and Research), has the objective of improving the life conditions of diabetic patients and relieving them of the burden of their pathology, through support from expert nurses and the provision of modern means facilitating their autonomy. The CERITD, under the leadership of Doctor Charpentier in collaboration with the South Francilien Hospital (CHSF), is engaged in a process of translational research on diabetes to develop new resources, particularly of telemedicine, allowing diabetic patients to self-monitor and to have the information necessary for the adjustment of their treatment. One of the first results of this research is a communicating medical device available on smartphones[42].

41 This is confirmed by recent studies on the evolution of the quality of life of people with kidney failure and transplant recipients: "The results of these QOL studies [quality of life] give arguments in favor of transplant to clinicians. The latter could tell their patients that, when the transplant is feasible and comparable to associated pathologies, life after transplant is better, particularly because of an increase in their physical capacities, less physical limitations and an improvement of their social relations", [SPE 14].

42 It concerns the Diabéo solution presented in Box 4.1. Source: www.ceritd.fr. Page consulted on May 12, 2015.

## 6.4.2. *"Calydial" association*[43]

### 6.4.2.1. *Readers of blood sugar, independent prickers and insulin pumps*

In diabetology, medical devices[44] are classified into two categories:

– *glycemic monitoring systems*: blood sugar readers, electrodes/strips, self-pricking devices, urinary strips, glycemic holters, automata and laboratory reagents[45];

– *the insulin administration systems*: the injector pens, needles for injection pens, rechargeable external insulin pumps (see templates below used in the treatment of independent insulin), implantable pumps, insulin syringes and insulin inhalers.

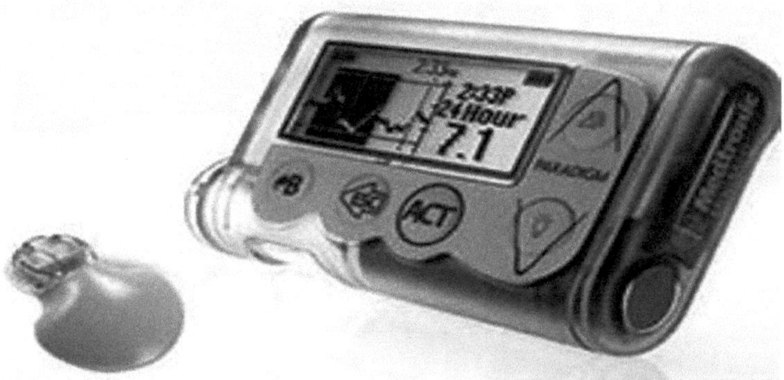


**Figure 6.5.** *Insulin pump (implanted or external)*[46]

43 See the description of the association in Chapter 6, p. 157.
44 "A medical device is an instrument, apparatus, equipment or even a software intended by its manufacturer to be used by humans for purposes such as diagnosis, prevention, control, treatment, mitigation of a disease or of an injury". Directive 93/42/EEC of the Council of the European Union from June 14, 1993 on medical devices. http://EUR-Lex.europa.EU/legal-content/FR/txt/?URI=CELEX%3A31993L0042. Page consulted on June 1, 2015.
45 Continuous Blood Monitoring (la Measure de la Glycémie en Continu – CGM), because of a sensor placed on the skin, gives a more comprehensive view highlighting hyper- and hypoglycemia than the rate of HbA1c and capillary glycemia, measured with a blood sugar meter, which it cannot always detect.
46 VEO 554/754 Medtronic paradigm model. Courtesy of Medtronic. The other big brand positioned in the market of computerized insulin pumps in France is Biotronik.

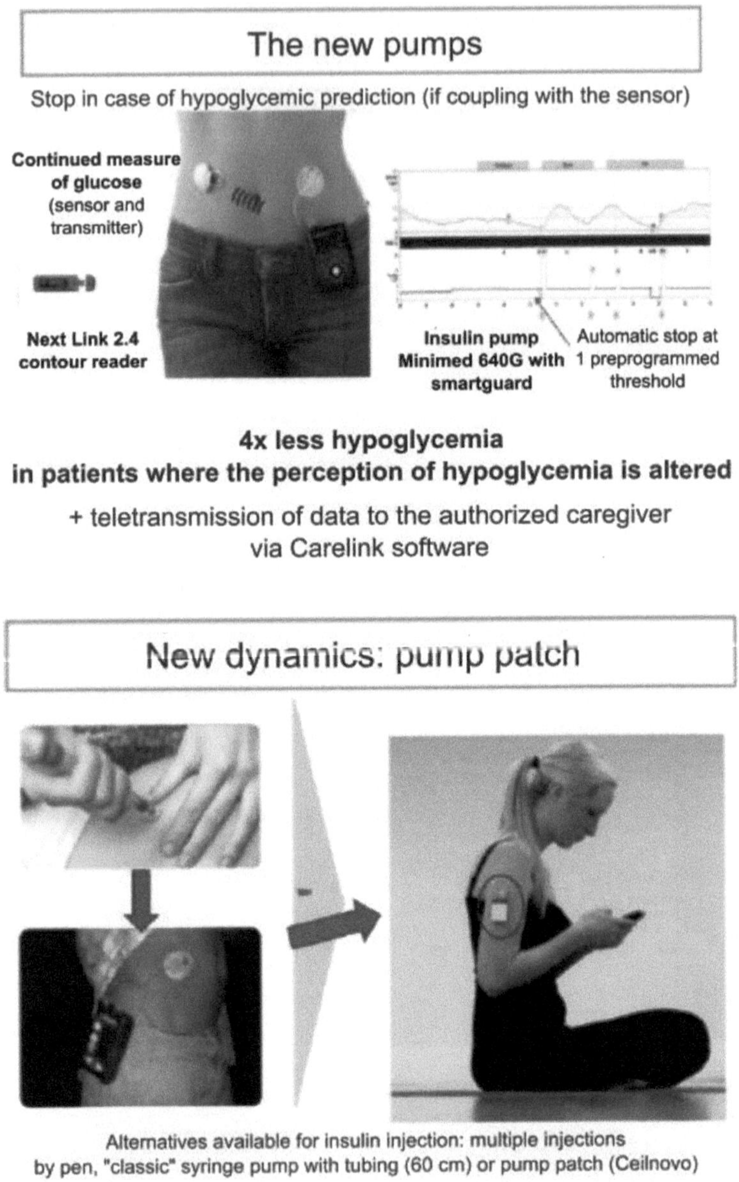


**Figure 6.6.** *Pump patch and external insulin pumps*[47]

47 http://ceritd.fr/la_pleniere/. Page consulted on December 12, 2014.

### 6.4.3. *Standard protocol*[48]

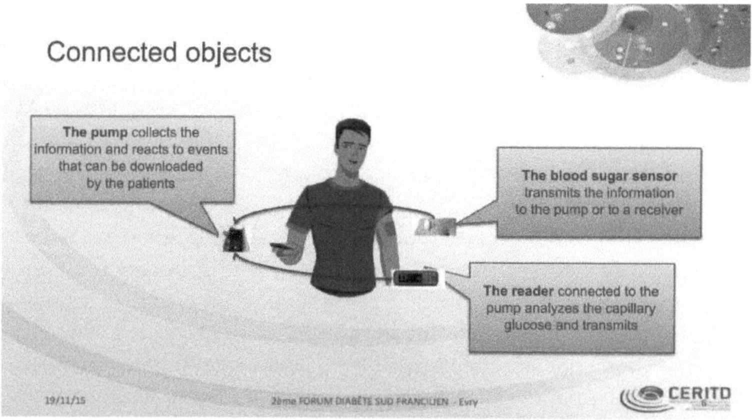


**Figure 6.7.** *Description of remote diabetes monitoring*

To perform a self-test, the patient shoots their fingertip using a finger pricker consisting of one or more reinforced lancets for single use (needle mounted on a base) on a support, with the appearance of a pen. The patient then applies the drop of blood on a test strip which they insert in the meter: the blood on the test strip triggers a chemical reaction that results in a change in color proportional to the value of blood sugar. This color is interpreted by the reader that then displays this value.

There is another type of reader, working not with strips but with electrodes: blood deposited on an electrode creates an electrochemical reaction that generates a microcurrent interpreted by the reader. Whatever the drive used, the patient has, then, the opportunity to see the results in their self-monitoring blood sugar diary in order to monitor blood sugar in the best manner possible and, more widely, their pathology. Note that recent blood sugar readers are thus connected as they allow, thanks to software, the directly downloading of the values of blood sugar on a computer and to statistically analyze changes in the patient's blood sugar.

**Box 6.6.** *Description of a remote monitoring standard protocol of a diabetic patient [SNI 14, pp. 6–7]*

48 Figures 6.7 and 6.8 presented in this paragraph are extracted from diabetologists Sylvia Frank and Gilles Lepape's presentation, entitled "Les nouveaux objets connectés des diabétiques" proposed during the 2nd Forum of South Francilien diabetes, November 19, 2015. http://ceritd.fr/les_ateliers/. Page consulted on February 1, 2016.

As part of their telemonitoring, the patient under home insulin therapy also indicates daily on the Diabéo application, apTelecare or other downloaded software on their smartphone or electronic tablet their blood sugar but also other data such as weight, level of physical activity and dietary details. Because of these data, these software companions, which have been previously configured following the doctor's prescription, can inform the medical team in real time. In the case of the Diabéo software, a message displayed on the smartphone to the patient shows them that their insulin at breakfast was sufficient but that it will take another dose at a time specified in advance. Because of their smartphone, the patient under independent insulin may send, via the internet, all of their data to the healthcare team who will respond quickly if it finds that the patient is found in a hypo or hyperglycemia state[49]. To date, care devices (blood sugar meter, insulin pump) are paired with a communicating smartphone or tablet type system. Tomorrow, a said "artificial" pancreas will combine these two functions and will become a powerful hybrid medical device: both device (implanted) care and communicating object used to transmit data from the patient directly to the healthcare team, without going through a communicating object (see Figure 6.8).

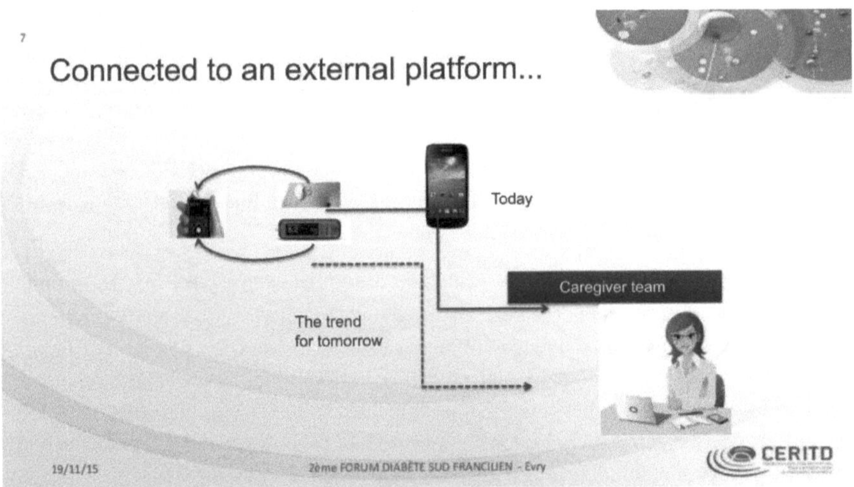


**Figure 6.8.** *Principle of telemonitoring in the context of home insulin therapy*

### 6.4.4. *Interviewees (and collected testimonies)*

Two patients with type 2 diabetes under independent home daily insulin, followed by Calydial, agreed to the proposed interview. In addition, the testimonies

49 For a detailed description of this telemonitoring protocol, see also section 5.4.

of three telemonitored patients with type 1 diabetes were collected: three men and two women for an average of 64 years of age. Table 6.5 lists the data collected (and anonymized) of interviewees under independent home insulin to study.

| **Patients** | **Age** | **Activity** | **Date of diagnosis** | **Monitoring center/start of home insulin therapy** | **Insulin pump model – care system** | **Mode of tele-monitoring** |
|---|---|---|---|---|---|---|
| Mr. R | 59 | No professional activity | Hereditary diabetes (1992) | Louis Pradel de Bron Hospital (69) (2015) | ANIMAS VIBE pump | Data stored on the pumps' memory card |
| Mr. M | 80 | Retired | 1990 | Portes du Sud Vénissieux Clinic (69) (2013) | Injection pen | apTelecare (Calydial) |
| Mr. P | 65 | Retired | 1995 | Montpellier Hospital | Insulin pump | Diabéo |
| Mrs. R | 66 | Full-time professional | 1960 | Montpellier Hospital | Insulin pump | Diabéo |
| Mrs. I | 49 | Part-time professional | 1981 | Montpellier Hospital | Insulin pump | Diabéo |

**Table 6.5.** *User data of insulin pumps concerned by the study*

### 6.4.5. *Opinions of people under independent home insulin therapy*

The typology of the impacts of the data used here, for people under independent home insulin therapy, is the same as that used previously for defibrillator carriers and people following independent dialysis.

| I) Positive *physical* impacts |
| --- |
| 1) "The sensor allows me to avoid repeated pricks" (Mrs. I). |
| 2) "The syringe involves four injections per day. The pump: it is a shot every two days" (Mr. R). |
| 3) "I am not obliged to wear the pump permanently" (Mr. P). |
| 4) "The pump is more convenient and more discreet [than the pen or the syringe]" (Mrs. I). |
| 5) "I can eat everything thanks to computerized tracking" (Mrs. R). |

| I) Negative *physical* impacts |
| --- |
| 1) The pump can be physically uncomfortable in certain situations. |
| 2) Replacement of the catheter(s) can be constraining. |
| 3) One's equipment must be constantly protected (pump, smartphone, reader, etc.). |

| II) Positive *temporal* impacts |
| --- |
| 1) "No change in daily activities since the use of the pump" (Mr. R). |
| 2) Reduction in the number of visits to the hospital and the diabetologist. |
| 3) "I can use my pump to break up the bolus during meals" (Mrs. I). |
| 4) The data collected by the pump and/or the blood sugar monitor are transmitted in real time to Diabéo (Mr. P). |
| 5) "I have a normal life. Before using the pump, I had to eat at a set time" (Mrs. R). |
| 6) "The pump fits my lifestyle and not vice versa" (Mr. P)[50]. |
| 7) "It offers great flexibility, I can better manage the insulin dose to send and choose my meal times" (Mrs. R). |

50 The pump is set to 24 hours: a base bolus (unit of insulin) defined by the diabetologist to cover the insulin needs on the day plus the bolus before each meal. The base bolus combination + daily bolus allows us to ensure insulin between the end of the night and breakfast needs, which was not necessarily possible under a protocol based on the use of an injection pen (risk in this case is higher in hypo- or hyperglycemia).

| III) Positive *psycho-emotional* impacts |
|---|
| 1) "The insulin pump provides a certain comfort" (Mrs. I).<br>2) "The pump: it's my friend" (Mr. P).<br>3) A sense of security and comfort is provided by the Calydial and Diabéo systems thanks to:<br>- the availability of nurses and doctors, including in the event of an emergency;<br>- psychological support provided by nurses;<br>- the electronic form to enter standard data: weight, change in catheter, blood sugar levels, physical activities, etc.;<br>- the monthly evaluation on the use of equipment and the difficulties encountered;<br>- the ability to manage one's own data. |

| III) Negative *psycho-emotional* impacts |
|---|
| 1) Daily use of the insulin pump and telemonitoring system may be constraining.<br>2) Support from loved ones can cause a transfer of anxiety.<br>3) The perspective of others can be difficult to handle sometimes. |

| IV) Positive impacts of communicating objects |
|---|
| 1) The use of a connected scale at home to track daily weight, associated with a therapeutic education, can increase the feeling of security.<br>2) The software system allows us to know, in real time, the dose of insulin that needs to be injected, which is reassuring.<br>3) One can do the bolus more quickly and more quietly by means of the connected objects. This gives a feeling of comfort (unlike the previous solutions: pens, syringes, etc.). |

| IV) Negative impacts of communicating objects |
|---|
| 1) Managing care at home in independent mode involves solving some stress-inducing problems yourself: possible machine malfunction (insulin pump) or software problems. |

| |
|---|
| 2) It is important to ensure that you have a mobile phone connection and internet access everywhere you go. |

| V) Areas for improvement in telemonitoring of independent insulin therapy | |
|---|---|
| Telecare technologies | Insulin pump:<br><br>Develop a "full" care process, i.e. automated data collection and sending of insulin by the pump without going through a manual entry of data (principle of the artificial pancreas). Increase the size of the reservoir (to space out the refills).<br><br>Sensor*:*<br><br>Offer care via social security in terms of the cost of the sensor placed on the arm (approximately 120 euros per month). |
| Therapeutic education | Optimize information campaigns on the insulin pump: there is a certain lack of information (many patients are unaware that pumps exist and they have a right to them). |
| **VI) Attitudes and opinions of people under autonomous independent insulin therapy** | |
| 1) Take a step back and learn to manage stress inherent to the illness.<br><br>2) Learn how to manage the perspective of others (during bolus and/or when the pump is visible).<br><br>3) Learn to master objects (pump, smartphone, software, etc.).<br><br>4) Learn how to manage one's diet and physical activity. | |

## 6.5. Recognizing the patient as a human and co-caregiver

The interviews conducted with chronic patients show that the new physiological order imposed by the disease triggers, as Catherine Tourette-Turgis, specialist of education therapy and founder of the University of Patients (UMPC – Sorbonne University) stresses, a redeployment of the subject:

> *Making an experience of the disease causes a series of questions that bear on one's own life, on life in general and the world. We cannot ignore the fact that we cannot ensure our biological sustainability. It is not the disease that makes the subject develop, it is the fact that the human being is in perpetual development, so that on the occasion of the disease, the patient uses tools and mobilizes resources because they are facing a new physiological order which forces them to invent a new pace of life as much at the physiological as the psychological and social levels. In this sense the disease is the opportunity for*

> *the subject to redeploy themself, to be transformed by their experience, and to transform it in turn* [TOU 15, p. 90].

This redeployment is facilitated today by care devices and communicating objects that offer the possibility for chronic patients to maintain or regain a social and/or professional activity. The telecare technologies, presented in this chapter, transform, by reversing it, the relationship between the temporality of the disease and the pace of life: they enable the carriers of defibrillators, dialysis patients, or those under independent insulin, to adapt telecare to their way of life, and not vice versa[51]. These technologies, therefore, optimize patient autonomy on a daily basis. The experience of telecare is, from the point of view of patients surveyed, an experience of a *controlled temporality* of their disease, which allows them to envisage their futures.

However, this redeployment of the subject is carried out *willy-nilly* because of the intrusiveness of telecare technology. Some patients seem to accept the implanted prostheses (defibrillators) or those that are detachable (computerized insulin pumps), as a "bad for good" or apply to forget them by overinvestment in the professional sphere; others see these devices as "guardian angels", or even "members" of the family (the technical objects go also on vacation, etc.). The intrusiveness of home care technology and communicating medical objects that are paired, although inevitable, seems *amenable.* The body's acceptance of prosthetics and care objects may however be facilitated. How?

The opinions of patients concerning their discomfort and/or pain caused by care devices, for example, must be heard. It is not enough to boast the technical-therapeutic merits of the "machines": we now need to accompany this discourse by a real consideration of the impacts of physical and psycho-social treatment of chronic diseases through machines. This observation, made by patients, deserves to be more widely shared by the telemedical community that serves them (practitioners, health engineer, nurses, psychologists, caregivers). Optimizing the management of patient's feelings of discomfort and pain would be a first step toward recognition as a human being or, if one prefers, as a feeling being.

The feeling of discomfort and pain, which must be heard by a psychologist, is both singular and intimate. Therefore, how can its care be generalized? Here again, the patient opinions deserve to be heard. The participation of expert patients during

51 This observation is shared, among other healthcare professionals, by nephrologists Hafedh Fessi and Pierre Ronco of the Tenon Hospital: "Each decade saw a breakthrough that revolutionized hemodialysis and this decade is that of home hemodialysis with new machines. This technique has a humanist dimension, allowing the patient to envisage themselves in the future, with a life project not punctuated by the constraint imposed by schedules and travels to centers. It also has significant social and societal implications" [FES 16].

hospital therapeutic education sessions was often mentioned during the interviews. The first benefit of this participation, which can be seen in various contexts (meetings with patients, reflection workshops, information days), is therapeutic: the participation of expert patients can indeed induce free speech and encourage empathic listening that contributes to patient well-being. Peer integration also allows us to value *secular expert knowledge*, which can be constructed by distancing the patient from their illness, which each participant can profit from. The principle of *peer-friendly mentoring* provided by the expert patients for patients, in connection with a therapist, is now an expectation widely shared by those interviewed.

Practitioners' empathic listening and benevolent peer mentoring allows us to take into account the narratives and the solicitations of chronic patients, induced by the disease and the telecare experience, but can also give meaning to the person's sense of well-being, as anthropologist David Le Breton outlines, in terms of *medical knowledge* and *know-how* valued by the clinic:

> *In any professional exercise requiring the involvement of the Other, three data fields become entangled: knowledge, know-how and self-management skills. In medical practice knowledge is valued by the faculty, the know-how by the clinic, and self-management skills by the patient. At the risk of losing an essential therapeutic vector, medicine cannot be cut from one or another of these dimensions. In this condition it doesn't treat the diseased organ, but the suffering man, knowing that he is immersed in a culture that is unique (representations, values, etc.), a social (lifestyle, etc.) and relational condition (couple, family, etc.) (...) The doctor does not escape these characters, he is also a man immersed in a culture (medical culture, but also that of his own education), a social condition, a relational framework, a psychology and a story of one's own. Against the temptation to impose his vision of the world (...), it is important that he treats his patient in the shared sense of equal dignity and mutual recognition. Therefore the conditions for human medicine are met.* [LEB 05, p. 117]

Self-management skills – "the ability to produce actions and reactions adapted to the human and ecological environment"[52] – should no longer be inscribed at the margins of the medical knowledge and know-how valued by the clinic, as David Le Breton suggests here, because it refers to the *pace of life* adopted by patients under the effect of the disease and the use of telecare technology. Self-management skills, or pace of life, refers to the strategies implemented by the patient, either spontaneously or through positive mentoring of peers, to manage the

52 Source: https://en.wikipedia.org/wiki/Social_skills.

psycho-emotional impacts, but also temporal and physical impacts of telecare. Self-management skills refer to the patient's ability to deal with the strain of home protocol practices (time and frequency of daily care), the physical and material management of care devices (malfunction, cleaning, replacement of any consumables) and/or smart objects (connection, transmission and reception), and the specific constraints related to chronic disease that are causing the patient, to varying degrees according to the pathology, to be "in tune with death", as a defibrillator carrier confided. In short, self-management skills in the context of telecare are a form of adaptation: they result from the independence acquired by patients through telemonitoring.

## 6.6. Expanding on the recognition of self-management skills

For healthcare professionals and expert patients, recognition of the person's skills and the effects of telecare can be valuable support in the empathetic approach. Extending the support of people in terms of self-management skills allows the doctor to treat the patient with a shared sense of equal dignity and mutual recognition. This sense of fairness can recognize the person as a *subject*, not an *object*, of care and breaks with a traditional vertical patient–doctor relationship. The overhanging posture of the omniscient caregiver does not have to be because telecare brings actors together in a collaborative relationship where patients are now part of the medical team. One of the major issues of telecare, in which the doctor is led to reassess his or her perimeter skills, lies in acceptance of the patient as co-caregiver, i.e. as a person able to evolve the care and knowledge about the disease, by their very own experience.

Recognizing the patient as a human and co-caregiver also commits the medical team to recognize their legitimacy to give their opinion on the treatment options that are offered to them and expand the responsibility of care in terms of the person's psycho-affective dimension. As pointed out by the interviewees, one must be able to eliminate malaise (which undermines self-management skills), talk about the effects of the disease and discuss the difficulties at home. Overall, the management of this dimension of the person is still little developed in France, even if quality psychological support can be offered today in some hospitals.

The uses and the relative fragility of care devices (for example the probe of a defibrillator) are also part of the concerns of patients and require a tight network of players concerned by telecare. This human *and* psychological support around technical issues is essential to care and patients[53]. These actors cannot assume the

53 The manufacturers of care devices, especially in the case of independent home dialysis, ensure close monitoring of the uses, as well as telephone support considered as globally efficient by the interviewees.

psychological “costs” of telecare alone, and practitioners cannot evolve telecare, and thus improve the health and emotional state of patients, without addressing the issue. Therefore, self-management skills are also a matter of psychological support in terms of use.

The advanced technologization of protocols and the doctor–patient relationship lead the telecare actors to work closely, to forge a therapeutic alliance as efficient, in the medical sense of the term, as human, in the sense of the patient. Identifying the temporal and psycho-emotional effects of telecare involves grasping two elements in particular: skills, as an adaptation strategy, but also the feelings related to the *physical* experience of care devices and communicating telemonitoring objects. Physically appropriating these *avatars*[54] of the body as a defibrillator, a cycler or a computerized insulin pump require overcoming, over time, in the view of interviewees, a psychological barrier: accepting *the incorporation* of an external element – which can be considered an intruder – and the potential risks associated with its use. The commitment of the person at the meeting point of biology and technology is an intimate relationship that is therefore experienced differently by each individual. However, there are several points of convergence in the way in which patients (all diseases combined) grasp the multiple facets of telecare and feel the effects.

## 6.7. Sharing the feeling of telecare

The review of the specificity of pathologies, telecare objects and their uses brings about the doubly singular experience of patients: specific to each disease and each individual. Beyond the diversity of pathologies and experience (temporal, physical and psycho-emotional), it is however proposed here to highlight and to examine the main features of the convergence points in the feeling of the multiple aspects of telecare expressed by patients (autonomy, psychological support, data sharing, home protocol, relationship with the disease, relationship with the third party, stress factors, etc.).

### 6.7.1. *Autonomy*

The particular feature of telecare is that it is a form of alternative technological optimized care provided in conventional medicine. The person – excluding the cardiac patient – can choose the time of daily care based on their activities or, if they cannot do this – when dialysis should be performed three or four times a day at regular intervals, for example – they can however capitalize on *home* care. As has

54 Here, the term “avatar” is used in the sense of an object to ensure technical support, internal or external, of a bodily function: heart rate (internal), insulin production and blood purification (external).

already been mentioned, the reduction of the number of trips to a hospital is one of the main benefits of autonomy of telemonitored patients. Interviewees testify to the comfort provided by care at home – or in terms of mobility (for diabetics) – compared to the alternatives (care in specialized units or drug treatments). Autonomy contributes to people's well-being because they can organize care or follow-up, in the case of defibrillator carriers, within their private space, according to their activities, their social, family and professional life. It also must be stressed that autonomy is assessed to varying degrees depending on the pathology and telecare: greater for a defibrillator carrier than for a person in peritoneal home dialysis (subject to frequent care), for example.

Autonomy, more or less conferred to individuals through telecare, because it offers the possibility of care at home, or in terms of mobility (in the case of the person under independent insulin), presents another major advantage: it allows relatives and carers to become involved in care and thus contribute to the person's well-being, even if, in some cases, the proximity between the person and their relatives can be a cause for concern: the latter could indeed feel a sense of anxiety, likely to affect the person in certain situations of everyday life. The interviewees acknowledge that this aspect of home care is preferable to the relative indifference which could be witnessed in a clinical or hospital setting.

For true autonomy, a tight-knit organization must be constructed. The pooling of human and technical resources around the person, within healthcare institutions, is indeed one of the possible conditions of autonomy at the level of the patient: doctors, nurses and manufacturers of care devices must collaborate closely in the interest of patients to provide follow-up and the permanence necessary for a smooth running of protocol. As a co-caregiver, the person is also required to participate in the protocol: the sharing of tasks related to telecare therefore confers a sense of empowerment to the autonomy that is appreciated. The person is effectively in a position to act in the face of the disease, under the watchful eye of the medical team.

Note that the organization of genuine autonomy, in the case of the dependent elderly people, includes using autonomy grids that today provide the basis of social aid, in particular AGGIR (see Box 6.7).

> Developed during the year 1997, the AGGIR grid (Autonomie Gérontologique Groupe Iso-Ressources) allows us to evaluate the degree of loss of autonomy or the degree of physical and psychological dependence in the accomplishment of daily acts of the person and groups seniors in 6 iso-resource groups or GIRs. This grid serves as a reference for obtaining APA (Aide Personnalisée à l'Autonomie [Personalized Autonomy Aid]).

**Box 6.7.** *Description of the AGGIR grid*[55]

55 Source: http://www.lesmaisonsderetraite.fr/maisons-de-retraite/evaluer-l-autonomie.htm.

### 6.7.2. *Psychological support*

The wish to be accompanied psychologically manifests itself differently depending on the people and the pathologies, and the psychological support, under the protocol, is appreciated to various degrees. Indeed, expressions such as *support* or *counseling* are part of the medical lexicon, which is not that of the interviewees. They do not feel sick *psychologically* and wish to simply be heard when expressing their discomfort with the disease. Psychological nurses in some health facilities, trained in empathic listening, can fill this expectation of people (all diseases combined). However, for patients, empathic listening from a healthcare professional is distinguished, in terms of its more or less formal nature, from informal support provided by loved ones, even if the search for emotional balance is the same. Empathic listening and counseling provided to patients are however not offered systematically as part of telecare. The carriers of defibrillators, in particular, demonstrated a real lack in this area.

Outside the protocol, empathic listening may be provided by other patients. Interviewees share the same feeling toward listening to peers: it is considered to be both reassuring and authentic as it is nourished by personal, bodily and emotional experience of the disease (which is not the case of listening to the doctor, nurse, relatives or caregivers). In addition, as evidenced by recent studies, counseling provided by a person with the same pathology allows improved care for people. Such is the case for people with diabetes[56]. At the same time, Doctor Michel Varroud-Vial (HAS) reports clinical studies demonstrating the benefits of counseling for caregivers in terms of care optimization:

> *Many studies have tried to assess the impact of such programs on the state of patient health. For example, a randomized study has tested this type of support in the US compared to the usual follow-up of 2,135 patients with diabetes, congestive heart failure or HTN. Supporters were trained for 2 weeks on the disease, the course of care and behavioral techniques. Their goal was to motivate patients to achieve the objectives that were recommended to them. The impact was measured by the number of objectives (for example for diabetics the absence of smoking, a HbA1c' <8%, a PA <130/80 mmHg, LDL-Cholesterol <1 g/l and the realization of follow-up exams). Each supporter followed 120 patients. On average, per patient, 2 one-on-one meetings, face-to-face in the doctor waiting rooms, and 5 telephone interviews have been carried out. After 12 months, patients helped reached more targets than the control group (82.6% vs. 79.1%;) and reduced the number of non-achieved objectives to 12.6%*

56 See, in particular, [ADA 13].

*vs. 30.1%. They also had less recourse to hospitalization (p = 0.05). The authors conclude that support by one health non-professional, integrated within the care team, seems to be an effective and inexpensive way to improve the results of caring for patients with chronic diseases*[57].

### 6.7.3. *Information sharing*

The data exchanged between the person and the medical team, because of telecare communicating objects (tablet, smartphone, transmission box), are shared knowledge that do not exclude the human dimension. By means of this common knowledge, it is possible, in effect, to build a joint effort: telecare, as has already been mentioned, is a practice designed for specific health objectives. These are known and negotiated by the patient and their doctor, remotely. A relationship in which the patient and doctor are co-responsible for (tele-) care can therefore be built. However, this co-partnership, as evidenced by the experience of the interviewees (all diseases combined), results when it is settled in the long term to a safe approximation of the doctor and the person as well as a common reality of the disease. Thus, a more relational dimension of care, corresponding to the expectations of the people and rewarding jointly the personal dimension of the data on the disease, as Gérard Dubey outlines, can be established through telecare:

> *What is expected is a relocation or re-customization of medical information; that disease is no longer being treated according to an objective point of view, but from the perspective of the patient's culture, singular and biographical elements, traces from which we identify in the world; relating to and interacting with each other. Also from the angle of personal interaction with the caregivers. Knowledge of the disease that therefore develops in the relationship and refers us to a social definition of health, more qualitative, but in which physicians, as practitioners, can also relate, in reality never being resolved to abandon this meaning of their profession. What connects so-called "expert" knowledge and so-called "lay" knowledge, is therefore their social condition of production, the fact that they constitute, precisely, a common reality*[58].

The reconciliation between the doctor and the person through the sharing of data is not just about customizing medical information: it is about making the doctor's

57 Source: http://www.has-sante.fr/portail/jcms/c_1701638/fr/association-francaise-des-diabetiques-afd-formation-de-patients-experts-pour-accompagner-les-malades.
58 [DUB 15]. See also [DUB 14].

expert knowledge and the patient's lay knowledge compatible and valuing the feeling of the person as well as their active involvement in the telemedicine protocol.

### 6.7.4. *Protocol at home*

For certain people (mainly those in independent dialysis)[59], a "hospital at home" can be seen as a social place. All actors of telecare and other healthcare professionals are found here. The private sphere becomes, as has been previously mentioned, a place of transformation and space for a new life. This transformation is that of the assisted patient (before their support at home) as an independent patient who can "find", because of telecare and according to their pathology, a certain freedom: to be able to work again, develop social activities, go on holiday, reinvest in social and family life. The "privatization" of the care protocol is therefore not only a material comfort but also provides a comfort and emotional stability highlighted by the people.

### 6.7.5. *Relationship to the disease*

Chronic disease, subject to study, influences people's survival[60]. The challenge of telecare is not the cure, but how the individual can remain in the best state of health possible. The relationship to the disease thus builds dynamic *resistance* of the person – vis-à-vis against possible complications – which is for the long term. The effects of the disease however must be contained daily by real-time monitoring (defibrillator carriers) and constant measuring of health (dialysis patients and diabetic patients) as part of telecare. The tension felt by the person between a long period with the disease and a short period of telecare changes the person's relationship with their disease: it is no longer simply an object of care but a new parameter of existence, an event that should *routinize* because of telecare in order to preserve social, professional and family lives. Interviewees share this objective of routinization of care, constitutive of a new look at themselves and the disease. The pathology is less of a concern in these conditions: it is put under control, because of telecare, to enable people to lead a more independent life and envisage their future lives.

Beyond the individual experience of the illness and the specificities of care/telemonitoring particular to each pathology, there are therefore convergence points in the feeling of the disease and telecare expressed by the interviewees.

---

59 Defibrillator carriers and diabetic patients are in effect not compelled to be at home and can enjoy greater autonomy thanks to the specificity of their equipment and their care and/or telefollowing.
60 See Figure 4.1.

Subject to the stress of incurability, always present, to the anxiety-provoking inevitability of care and/or remote monitoring, they can acquire a significant autonomy (but relative according to the pathology), find a reassuring emotional balance, but precarious (as subject to the whims of the disease), an empathic alleviating form of listening, though unfairly granted according to pathologies, as well as a certain satisfaction in the eventual reconciliation that can take place (according to the pathology) with the doctor, because of telecare. Management of care remotely, from a distance, and the impersonality that it assumes must, however, be supplemented imperatively, in the view of interviewees, by emotional support. We confirm here the central role of affectivity in the experience of the illness already evaluated in a phenomenological perspective[61], and the therapeutic interest of its management by telecare. What is commonly called the humanization of protocol – i.e. increasing attention to the relational dimension of care – is actually, in the case of managing chronic diseases remotely, a process of recognizing the subjectivizing value of affectivity and its therapeutic value in a perspective of wellness including co-construction carried out by the person, the doctor and the medical team. In a nutshell, the feeling of telecare shows the importance that should be given to the *challenging* nature of chronic disease.

## 6.8. Reconsidering the order of the person

Studies on the quality of life of chronic patients engaged in home telemonitoring protocols often point to the main inherent risk in the technical care relationship: not perceiving the person in terms of their cognition, in all their singularity but as a technical factor. The summary of patient feelings, both positive and negative, allows us here to shed light on the real possibility of taking into account, or even rehabilitating, the order of the person in the context of telecare. For the person, telecare does not simply help them avoid pain and disease: it allows them to spend time with their family, enjoy hobbies, voluntarily putting the disease at a distance, not forgetting it but taking a step back from it and refocusing on life.

At the same time, the recognition of the patient as a person allows the medical team to differentiate *the emotional state*, as a perceived and recognized objective phenomenon, with its physiological and psychological specificities, from the emotional experience that represents what the individual, plagued by an emotion, feels and of which they may become aware of[62]. The emotional experience, which involves a cognitive capacity to apprehend the meaning of the situation with which the person is engaged, the perception of the physical manifestations, recognition of the emotion itself, can currently be supported, as the experience of patients surveyed by empathic listening of caregivers in the context of telecare shows.

61 See section 3.7.
62 See [LEW 00].

What is usually called the psychological dimension of the patient and is actually, in part at least, the emotional experience of the person has always been deliberately excluded, for reasons that are both historical and ethical, from the doctor–patient relationship because it refers to the person. In the organized model of medicine – the one that is the object of an ordinance, meaning "an emanated prescription from a higher authority" (Littré) – the patient as a person is absent, the content of the speech will focus on a defined illness by natural laws. The order of the person, historically denied, calls for a reclassification in the context of telecare because the patient now accesses the status of co-caregiver and thereby acquires even a new legitimacy in the relationship of care that requires a recognition of their experience, both psycho-emotionally and techno-therapeutically. This recognition can be done upstream, at the prevention stage as well as downstream, after treatment. The debate around recognizing patient experience implies therefore questioning the ordered model of medicine that has regulated consultation since the appearance of Greek medicine in the fourth and fifth centuries centered around the three steps described by Plato: examination, diagnosis and treatment[63]. This model has of course evolved but it always involves a competent practitioner making decisions and a patient submitting to them. In short, the requalification of the order of the person in the telecare context implies changing the two main concepts of the ordered model of medicine: first, that of the disease, reducible to physical and chemical laws, and the second related to the carrier of the disease, the neutral care object. This development points to *perception*, meaning to the feeling person beyond the patient but also, ultimately, to a therapeutic alliance of a new kind.

## 6.9. Toward a biopsychosocial telecare model

DEFINITION 6.3.– *Empowerment* means "the process by which an individual and/or a group acquires the means to strengthen its capacity for action allowing access to individual and collective power". "[This term] articulates two dimensions, power, which is the root of the word, and the process of learning to get there"[64].

The *empowered* person in the context of telecare can be supported in their emotional dimension, as evidenced by some current experiences, without this care being acquired for all patients. The interviewees effectively confirm the deficit of compassionate listening often referred to by their peers. However, telecare is not a device of impersonal medicine. It does not consider the patient as an object of care. It certainly requires the doctor to be of emotional neutrality, not altering their

63 See Plato, "Laws", 720, c-d and "The Symposium", 586, in [PLA 97]. Plato, *Œuvres complètes*, Flammarion, Paris, 2008.
64 [BIE 11]. See also [BAR 15].

judgment, but seeing that they have an *obligation to manage emotions*. Initiatives are taken by doctors and patient associations to evolve a more humane form of patient care. Health insurance is not absent from this development (see Box 6.8).

> The mission of Sophia, a support service for health insurance, is to help people learn more about their disease and to adapt their habits to improve their quality of life and reduce the risk of complications. By bridging together the recommendations of the attending physician, Sophia offers support, information and personalized advice, adapted to the situation and needs of each patient. The person can also share the experience of other patients and healthcare professionals. The Sophia service already supports more than 630,000 members. It is offered in all of France for people with diabetes, and 18 "départements" to people with asthma thus far. The Sophia service, however, does not offer psychological support, but does offer telephone *motivational interviewing* by nurse counselors in the therapeutic follow-up framework (targeted goals: improvement of observance, awareness of the disease, etc.).[65]

**Box 6.8.** *Description of Sophia, a support service for health insurance*

Evidence also shows that there are *compassionate* healthcare professionals (nurses, doctors, surgeons) to listen to their patients in some public and private health facilities. As we have seen, some areas are better equipped than others in human resources and equipment, and psychological support is more or less ensured according to pathologies.

Outside of a strictly rational health model (biomedical model), telecare brings telemedicine into a transitional phase. Today, this is changing under the injunction of patients and patient associations and some health professionals toward a broader recognition of the temporal regime and the psycho-affective dimension of the person. This development transforms the perspective traditionally conveyed by the patient in conventional medicine; a broader view of the patient as a trying person, a bearer of values, beliefs and possessing a personal and committed history in a community, is encouraged and valued. What patients say through their testimonies is that the biomedical medicine model needs to evolve to meet the needs of patients in terms of moral and physical suffering. What the patients say, in essence, is that it is essential to continue advancing telemedicine from a biomedical model toward a *biopsychosocial model.* The evolution of contemporary telemedicine includes driving a change toward new best practices, and a conception of the *patient-as-person*, similar to the transformation that takes place at this level, as Xavier Guchet outlines, in the field of personalized medicine today:

65 Source: https://www.ameli-sophia.fr.

> *Undeniably, public health players have to meet the expectations of citizens who want medical and healthcare facilities that consider them first and foremost, not as objectified bodies, but as people facing the disease painfully with psychological, social, familial or even professional problems.* [GUC 16, pp. 415–416]

Therefore, the biopsychosocial telecare model means a special techno-medical model in which the biological and the psychological nature of the person can be reconciled and jointly supported. The social dimension of this model is twofold: it is in a doctor–patient relationship humanized through independence acquired by the person and in a sociability found through telecare. In addition, the biopsychosocial model is based on advanced technologies that perform a dual function. On the one hand, mediation provided by the objects is symbolic – these are carriers of sense, values, emotions – and make it possible, as discussed in Chapter 7, for the projection of the person – in the sense of building a project of oneself. On the other hand, mediation provided by telecare within this biopsychosocial model is "concrete" because it weighs on the different registers of temporality and affectivity of the telemonitored people.

The choice was therefore made to consider, in the third part of this book, the specificities of the interaction of people with telecare objects (implantable, detachable prostheses and communicating objects). The question of the temporal economy of this technical mediation, both symbolic and concrete, as we will see, poses, in a final instance, the moral question of the status of telecare objects as chrono-technologies transforming the relation to ourselves, to disease and the temporality of the person.

# PART 3

# Toward an Ethics of "Time-sensitive" Telecare

# Introduction to Part 3

Medico-economic studies on telecare primarily evaluate the therapeutic effectiveness and the cost–benefit ratio of existing platforms[1] and, incidentally, the potential violation risks of confidential patient data and medical confidentiality. However, it is appropriate to distinguish, on the one hand, the cost–benefit relationship for the medical institution, and, on the other hand, the cost–benefit ratio which expresses itself, for the patient, in terms of the medical service administered. It is not a matter of denying the value of this medical service, but simply enlightening its various facets by adopting the patients' point of view, including on a technical level. In fact, the patient's health benefits are not exempt from risks due to the use of computerized devices and objects connected on a daily basis. Beyond the technical aspects, and taking into account the value of the cost–benefit ratio of telecare, it is essential to assess the patient's subjective dimension, particularly in their interaction with digital health devices, whose care is becoming more critical today.

The testimonies collected from patients reveal the need to expand patient care to non-technological telecare aspects, including a psycho-emotional dimension. Therefore, the thoughts proposed in this third part pursue a double objective: on the one hand, to grasp the preventive dimension of telecare through the examination of the uses of care devices and communicating objects that shape the perception of the patient's future; this perception being influenced biologically by the affects, as evidenced by cognitive sciences; and, on the other hand, to identify the specificities and the ethical scope of relational technologies sensitive to the temporal regime *and* the psycho-affective dimension of the person.

The study of these two complementary facets of telecare, which has never been completed to date, aims, on the one hand, to inform people about the ethical status of

1 Concerning this point, see, for example, [PAR 10]. Also see [STH 06] and [ABD 11].

technologies "sensitive" to the temporality of the person, and of the disease – "chrono-technologies" reslotted here – in the framework of telemonitoring and the homecare of cardiac patients, and patients treated by autonomous dialysis or insulin therapy. To take care of the person, in this context, means to address the *objective* and *subjective* temporal care aspects and to support the disease and the person by taking advantage of technologies – not to subject the person allegedly "autonomized" and "empowered" to objectifying care techniques. One of the goals of this reflection is to to shed light on an enlarged concept of autonomy. It can be acquired jointly as we have seen, by therapeutic education *and* by experience of the disease and telecare.

Finally, the study seeks to question the concept of "time-sensitivity" from which an ethics of *telecare*, acknowledging its debt to medical ethics and care ethics[2], can be built to frame abiding practices of the temporal regime of the person whose therapeutic value will be highlighted here. Sensitive to the intimate time and chronology of the disease, and even sensitive to chronobiology, chrono-sensitive telecare is a relational ethic[3]. In other words, the sensitive experience of the disease and care, consistent with such an ethic, is an integral part of human "settings", which must today be integrated into telecare procedures. Technical mediation operated by the telecare technologies is therefore not only symbolic, as shown by the study of the affordances in relation to the project which may be carried out by the sick person (see Chapter 7 of this book). It is also concrete since it transforms the relationship with oneself, the disease, the temporality of the person and the doctor–patient relationship in a sense – that of the personal being – that can no longer be ignored by the telecare actors (see Chapter 8 of this book).

The ethics of time-sensitive telecare aims precisely at guiding telecare practices in the sense of a reconciliation of humanity and the technical nature of care.

2 See Chapter 2 of this book.

3 As we will see in Chapter 8, chrono-technologies have also acquired a moral value in the protocols, which is constitutive of a telecare ethics centered on the patient.

# 7

# Subjectivizing the Future: or, the "Patient-Project" Temporality

*The consciousness of time is always for us, an awareness of the use of the moments, it is always active, never passive, in short the conscienceness of our own duration is the consciousness of a progression of our intimate being.*

Gaston Bachelard

The communicating objects used for the telemonitoring of diabetes and kidney and heart failure at home (smartphone, e-tablet, sensor, implantable computerized devices, removable prosthetics, orthotics) allow practitioners to conduct periodical surveys of the blood glucose level, heart rate and life data (blood pressure, weight, temperature) which are *equally* part of the history of the disease. The function of these objects is to ensure quality follow-up on the person, anticipating any possible deterioration of their state of health. Concretely, communicating objects allow the cardiac patient to provide information about their pathology; the patient with kidney failure to collect and transmit, to the telepractitioner, the data allowing them to perform home peritoneal dialysis; and the diabetic patient to inject the dose of insulin required at the right time, according to need. Communicating objects are thus intended to collect and store the patient data, to promote their analysis in real time, to remotely track the care performed by the patient and, more generally, to anticipate the state of health of the telemonitored person.

These communicating objects produce *time-images* of the chronic disease, on a daily basis, i.e a series of instantaneous (life data entered at an instant *t*) constitutive to the field of the diagnosis, care, understanding and action *by anticipation* from the patient and doctor. This chronological reordering of the disease by interposed objects – a kind of "chronoscopy" of affection – and preventive quantification of body symptoms (electrocardiograms, increase in blood pressure or weight) favor a selective medical gaze. Essentially based on remotely gathered *data*, but no longer on data obtained in the one-on-one conference, this view operates on the diseased person a sort of anatomopathological reduction by spurring them to focus on the event nature of the disease (arrhythmias, hypoglycemia, hemorrhage) and physiological markers of *circadian coordination* (activity/rest, food intake, body temperature). However, this *techno-centered* look at the disease takes insufficient account of the proven influence of the psychological markers for the synchronization of biological clocks, such as emotions and psycho-socioenvironmental factors.

In the context of home telemonitoring, the psychological factors of the disease are usually supported by social or family caregivers. This can help the person to overcome the obstacles encountered in the use of smart communication objects and the implementation of the home care or communication with caregivers, and to ensure the overall emotional balance of the person (elements linked to stress and anxiety). Caregivers can provide moral support to the patient, but the understanding of the psychological factors of the disease is not limited to empathic listening at home. In addition, the disease cannot simply be understood from a cause–effect perspective regardless of the emotional health of the person. In the work of caregiving, it is important to take into account the human being who "functions" as an "interconnected" system: bodily mechanisms as well as emotions and the perception of time are constitutive to the illness experience. In this context, and in view of recent advances in applied cognitive science, increased attention should be given to the links found between a patient's emotions and temporal perception[1].

In the context of telecare, the role of emotions and temporal perception indeed deserves to be acknowledged for two main reasons: on the one hand, the time experienced by the patient, which is the consciential time described by Gaston Bachelard in *La dialectique de la durée*[2], promotes the redesign of experiences, impressions and memory traces based on new experiences in guiding action. On the

1 Cognitive science agrees that our perception of time is highly "influenced" by our emotions, as highlighted, among others, by psychologists Ruth S. Ogden and David Moore: "In the literature on temporal perception, the fact that our experience of time is influenced by the emotions is a robust finding. The most consistent result relates to the fact that the effects of negative stimuli causing fear last longer than the neutral stimuli with an equivalent term", cited in [OGD 15].

2 In *La dialectique de la durée*, Bachelard constantly speaks of three subjective variants of time. They appear as a background in physics, and most significantly in psychology and musical art: time in which we are living (existential time), time that we feel (consciential time) and time that we shape (ideal time).

other hand, temporal perception and emotions are *feelings* that participate fully in self-knowledge and the construction of a self-image not only in the present, but also in the future. To fully understand the specific nature and scope of this double phenomenon, it is appropriate here:

– to shed light on the biological basis of emotions and the patient's temporal perception and their role in understanding the disease;

– to explain psychological time scales and show how the future, more than the past and the present, involves the person in a self-reflexive behavior in the context of telecare;

– to clear the epistemological meaning of the subjectivation process of the future, through the examination of the telepatient's anticipatory mechanisms, as a condition for the development of a project for themselves.

This triple analysis allows better understanding of the role played by anticipation in the patient's temporal consciousness. As a cognitive act and thought mechanism, anticipation appears as a movement that engages the patient's identity to project it into a space–time of potentialities. Anticipation allows us to operate, in particular, a passage from the registry of prevention, generating a temporo-existential break (to be a simple telemonitored "subject" who believes they have no future because it is already predetermined by the chronic disease), to the affirmation by the subject of their own destiny. Anticipation is a mechanism of thought, related to emotions, which is transmuted into an ability to sustainably ensure a link between the sense of a patient's future and *their* project; it helps to maintain the unity of the being in time. But is this singular property of anticipation, specific to the individual, supported by telecare?

In reality, the telemonitored person, as a *temporal being* involved in an individuation process, is not understood in their ability to assess their own future in the context of telemonitoring today. This phenomenon must be attributed to the impossibility of *objectively* measuring the therapeutic gain of the person's psycho-emotional feelings and their anticipation potential. Centered on the quantitative measure of the temporality of affection, the communicating objects are not intended to ensure the person's psychosocial well-being. These objects are certainly not simple transmitters/receivers of data, artifacts that influence, like our relationships with others, the existential quality of our experience of the possible world, but their functionality requires technical normativity. Does this mean that telemonitoring reduces the patient, by interposed objects, to an organized living matter, without changing their temporal perception of disease or sense of becoming?

The advanced hypothesis in this chapter, in a second analysis, is that objects by their affordances (relational dispositions), exert an indirect influence on the patient's emotions, temporal perception and future. The term *affordance* has a double meaning: "ability to do something" and "offer". A concept coined by psychologist James Gibson, *affordance* here, refers to the ability of an object to suggest its own use in relation to a subject[3]. It also concerns the intuitive use (or intuitive character) of an object.

By their ability to measure the temporality of the disease, objects can influence the patient's perception of time (symptom, care, protocol), their perception and production of rhythms (cardiac, circadian) and their perception of the temporal order and concurrency (synchronization with the telemonitoring system). The intuitive use of objects by the patient, that is, to seize the ability of an object to change their perception of time beyond its primary communicative function, can also contribute to development-oriented temporal representations of the future in the patient's *implicit reasoning*, i.e. anticipatory intuition (ability to act, represent the future), and their *explicit reasoning* (planning, anticipation, recovery of memories related to the disease). In one word, the objects allow the patient to potentially "coordinate" with the future. Objects thus participate in the movement of subjectivation, which leads the person beyond maintaining self-support assured by telemonitoring, to *redefining themselves as a project*.

In this dual movement of anticipation and individuation, objects however play less of a *structuring* than a *transducing* role, in the Simondonian sense of the term: the technical standard of anticipation that objects establish (prevent to better treat) enables, under the effect of affordances, the advent of a form of self-action by the patient, by which they may become aware of not being a substantial but a *transductive* individual: meaning not entirely contained within their current limits, out of phase with themselves, facing the past and the future, committed to a personal project through the communicating objects (anticipate living better). The two-fold issue that is evident behind this analysis is *in fine* that of the patient as *poiesis*, meaning an anthropotechnic construction, and as a *subjectivising conscienceness* of their own future.

DEFINITION 7.1.– Gilbert Simondon gives a definition of transduction as "operation", which is particularly appropriate for our intentions: "There is transduction when there is activity from a center of the being, structural and functional, and extending in various directions from this center, as if multiple dimensions are to appear around this center; transduction is the correlative

3 See also the elucidation of this term: Chapter 7, section 7.8.

and structures in a being in a state of preindividual tension, meaning a being who is more than a unified being and more than an identity, and who is not yet out of phase with themselves in multiple dimensions" [SIM 05, p. 33].

In summary, this chapter aims to shed some light on *the protention of the patient as an action project* subordinate to their use of communicating objects. In the psychology of anticipation, protention indicates the fact that a person is looking to the future. This clarification tends to show how the future unravels, as a persistent matter, within the issue of *authority to act* raised by the telemonitored person's self-acquired (temporal) knowledge, and enables us to measure the epistemological scope of the "patient-project" concept that emerges from this study.

## 7.1. Anatomy of anticipation

The review of the biological basis of emotions and the perception of time in humans, proposed in this section, is intended to highlight the epistemological value of *feeling* and emotions in self-knowledge and self-construction for the future. Thus, delimited feeling helps influence the person's subjective judgment and, in the case of a chronic patient, sharpen awareness of their techno-medical environment, their chronic illness and their future. In other words, identifying the emotions and feeling of the person, at these three levels, allows better understanding of the role of subjectivity and proactive mechanisms of the conscienceness that make self-projection in the future possible. Here, anticipation returns between the lines to an early detection of biological phenomena at the level of biomarkers. However, highlighting the biological bases of emotions and temporal perception that underlie the process of *subjectivation of the future*, based on the affects, requires raising, from the outset, several questions to guide the study.

QUESTION 7.1.– What do we call "biological or circadian rhythms"? How is the knowledge of these rhythms useful for understanding the evolution of the disease in the context of telemonitoring? What is the temporality regime of the telemonitored person? What are the specificities of their experiences (physical and psychological) of the chronic disease? What is the influence of the affects on the temporal perception of the disease[4]?

The answers provided in this chapter first benefit from the teachings of chronobiology. This science is concerned with biological rhythms and their effect on health, external elements that have an impact on them, their importance to health

4 Concerning this point, see, in particular, [ANG 97].

and elements that are susceptible to disturbing them. Chronobiology identifies three categories of rhythms that indeed offer a useful theoretical framework for the understanding of the patient's bodily experience: the ultradian, circadian and infradian rhythms. Ultradian rhythms are biological cycles occurring within 24 hours, like a heart rate that lasts less than a second (although it repeats in a loop). Circadian rhythms are biological rhythms that take place globally in a 24-hour period. We include, in particular, among these rhythms: changes in body temperature, sleep cycle or even the rate of hair growth. Finally, infradian rhythms are cycles produced beyond a 24-hour period. For example, they can occur every month, such as the menstrual cycle[5]. Here, we will focus mainly on circadian rhythms because of their influence on the state of health of individuals. The so-called circadian rhythm[6] regulates most of the biological and behavioral human functions.

## 7.2. The clocks that govern us

Diverse bodily functions such as blood pressure, hormone production, heart rate, cognitive capacities, mood and memory production are regulated by the circadian rhythm. Conversely, rhythm deregulation leads to sleep disorders and severe physiological disruption. More generally, chronobiology research shows that nearly all human biological functions are subject to this rhythm. The manifestations of this cyclical activity are numerous: thanks to the circadian clock, melatonin is secreted at the beginning of the night, sleep is deeper toward 2 o'clock in the morning, body temperature is lower early in the morning and higher at the end of the day, intestinal contractions decrease at night, arousal is maximum from mid-morning to late afternoon, memory is consolidated during nocturnal sleep, etc. (see Figure 7.1). Studies in chronobiology also show that isolation of humans for several weeks, in conditions of near-total darkness and without temporal reference, does not prevent the persistence of a cycle where rest and activity alternate on a 24-hour period. This persistence proves that the circadian rhythm is endogenous: it is generated by the body itself. The organism therefore has a set of synchronizers, more commonly known as the "internal clock", nestled in the center of the brain, which acts as a metronome for the body. A multitude of physiological functions ranging from gene expression to behavioral phenomena are therefore controlled by the biological clock (see Table 7.1). This synchronization between the different physiological functions allows the body to function optimally [ALB 12].

---

5 Source: www.sante-medecine.commentcamarche.net. Page consulted on March 22, 2015.
6 Circa: "close to", dian: "a day".

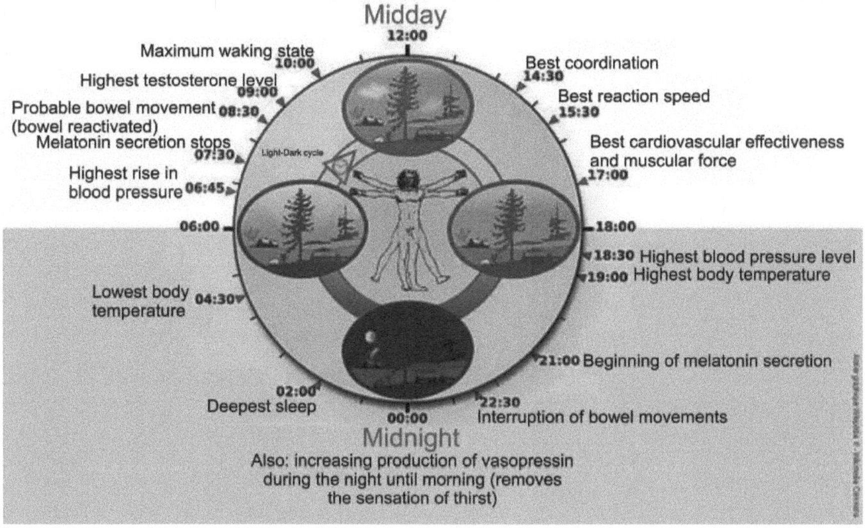


**Figure 7.1.** *Ultradian biological rhythms and human circadian rhythms*[7]

| Examples of variable circadian systems | |
|---|---|
| Cardiovascular | Heart rate<br>Blood volume<br>Blood pressure |
| Respiration | Respiratory rate<br>Airway resistance<br>Severity of asthma |
| Endocrine system | Growth hormone<br>Cortisol<br>Norepinephrine |

7 Source: https://en.wikipedia.org/wiki/Chronobiology. Page consulted on February 2, 2014.

| | |
|---|---|
| Gastrointestinal | Gastric acid secretion<br>Appetite<br>Digestion/Excretion |
| Thermoregulation | Internal body temperature<br>Sudation<br>Vasoconstriction |
| Immune | Production of antigens<br>Production of cytokines<br>Effectiveness of immunosuppressants |
| Vision | Retinal sensitivity<br>Saccades speeds<br>Pupillary reflex |
| Central nervous system | Cognitive performance<br>Melatonin secretion<br>Sensitivity to magnetic stimulation |
| Behavior | Sleep cycle<br>Athletic performance<br>Muscle strength/torque |
| Disease | Cancer progression<br>Heart attack<br>Treatment effectiveness |

**Table 7.1.** *Systems under circadian control in humans*[8]

Besides its endogenous dimension, corresponding to a genetically determined rhythm persisting after the elimination of periodical environmental information, all circadian rhythmicity also has an exogenous component that depends on the rhythms of environmental factors called synchronizers. Endogenous rhythms are adjusted in time or synchronized by time donors or synchronizers that are periodic environmental factors: day–night or photoperiod cycles, alternating rest–activity, meal times. Among periodic factors of the environment, light–dark cycles play a vital role in the synchronization of circadian rhythms in humans and exposure to a

8 Summarized and adapted from [TUR 99].

light pulse can shift the production peak (called phase) of a hormone (cortisol, melatonin, thyrotropine, etc.; see Box 7.1). Depending on the time of exposure, the phase will be advanced or delayed. Exposing a healthy subject to light (2,500 lux) for just three hours from 5 am to 8 am for six consecutive days causes a decrease in the plasma concentrations of cortisol and melatonin and a shift of the circadian rhythm phase of these two hormones (called "phase advance"). In a subject with good health in which the body lives in harmony with its environment, the circadian rhythms of the different functions are normally synchronized, which means that they have, with respect to each other, a specific phase relationship. This organization and temporal coordination function of biological rhythms is mainly performed by small structures in the brain made up of 10,000 neurons, i.e. suprachiasmatic nuclei (SCN). These structures receive, directly from the retina, the photoperiodic signal, which is a major synchronizer of circadian rhythms. Neurobiologist Claude Gronfier explains this phenomenon:

> *The light is captured at the level of the retina by a group of specific photoreceptor cells (melanopsin ganglion cells), related to the suprachiasmatic nuclei by a different nervous system involved in visual perception. The signal transmitted to the internal clock leads to remission at the cycle time to sync it in 24 hours. This same signal is also transmitted to other so-called "non visual" cerebral structures that are notably involved in regulating mood, memory, cognition, and sleep*[9].

If the SCN are harmed, the peripheral clocks are out of sync: a domestic phenomenon that is most notably observed during aging and in some chronic pathologies.

| |
|---|
| Cortisol (or hydrocortisone) is a steroid hormone secreted by the cortex of the adrenal gland from cholesterol. The thyrotropin-releasing hormone (TRH) is a peptide hormone produced by the hypothalamus that stimulates the secretion of thyrotropin (TSH) and prolactin by the anterior pituitary gland. Melatonin, a synchronizer under the influence of light, is a hormone whose secretion is typically circadian. Its production increases by the end of day just before sunset, contributing to falling asleep. It reaches its secretion peak between 2 and 4 o'clock in the morning (see also Figure 7.1). |

**Box 7.1.** *Hormones involved in circadian rhythms*[10]

9 Quote and data from [CHR 13]. Source: www.inserm.fr. Site consulted on February 2, 2014.

10 Source: http://en.wikipedia.org/wiki/Cortisol. Page consulted on February 2, 2014.

## 7.3. Stressors

A desynchronization usually refers to a state where two (or more) rhythmic variables previously synchronized have ceased to present the same frequency and/or acrophase relationships and show different temporal relationships of usual relationships[11]. Desynchronization may be external, depending on changes in the environment, and is found, for example, on a transmeridian flight of five zones (jet-lag phenomena) or in a work shift. Internal desynchronization does not depend on environmental factors. It is found in aging subjects or in a number of diseases such as depression or certain cancers. This desynchronization comes, whether external or internal, with a set of atypical signs such as fatigue, sleep quality, mood or appetite disorders. These disorders can be corrected by a treatment of desynchronization: intense light (seasonal depression) or administration of melatonin (in the case of jet lag).

Stress induced by external factors, such as noise, temperature and especially light, can disrupt the circadian rhythm by delaying, for example, the secretion of melatonin. However, to date, no communicating objects include stress measurement or corrective functions induced by external factors in the protocols of telemonitoring chronic illnesses (diabetes, respiratory, cardiac and renal failure). However, the stress associated with changes in light, for example, has an impact found in some patients, particularly those with diabetes or Alzheimer's disease. The decrease in light late in the day, for example, causes a stress called "twilight" that can trigger in these patients the production of negative emotions (neuroticism) and a series of physiological effects that may affect circadian rhythm. Current light therapy techniques used to address this type of stress which is potentially aggravating to these patients are not integrated in the protocols of telemonitoring of chronic patients. The perception of time, which is based on the degree of brightness, is nevertheless a determining factor in the patient's physical experience, telemonitored or not. Stress induced by external factors can also change the sleep patterns of the chronic patient. Knowledge of the circadian rhythm, which is essential to better reorganize one's sleep, also called "chronotype", can be gained using a morning–evening test like the one designed by Jim Horne and Olov Östberg in 1975, validated by multiple sleep research centers and which today remains the reference[12]. To evaluate desynchronization of the biological rhythm, there are also two types of psychological and physiological markers (see Box 7.2).

---

11 Acrophase is the position on the axis of time with the maximum variation in a temporal biological rhythm.

12 Originally published in the *International Journal of Chronobiology*, a 19-question test from Jim Horne and Olov Östberg allowing one, for the first time, to define one's chronotype: morning, evening or neutral (see [HOR 75]).

| Psychological markers | Physiological markers |
|---|---|
| Stress (psychological) | Activity rest |
| Emotions<br>*(neuroticism: chronic experience of negative emotions)*<br>(Mental compensatory strategy, known as "coping") | Food intake |
| | Body temperature |

**Box 7.2.** *Main markers of circadian coordination*

The recognition of the temporal structure of the organism has implications in clinical medicine in terms of establishment of the chronobiological reference values, early detection of risk situations and therapeutic approaches that take into account the temporal variations of the desirable or undesirable effects of a drug. But the recognition of the temporal structure of the organization also allows us to highlight the internal and external circardian synchronization and desynchronizaton markers, mainly stress and emotions, that act on how the patient sees their affection and negotiates their interactions at home with communicating objects, the telemonitoring system and the telepractitioner. Stress and emotions are in effect the principal markers of the telemonitored person's feelings that act, as we will see, on the singular patient–object interaction in the context of telemonitoring. To better appreciate the role of emotions in this context of self-knowledge, however, it is important to sketch the outlines of the temporal ecosystem within which this feeling can be experienced. The particularity of the patient's real-life experience and emotional states is in effect their anchor in multiple temporalities of home medical telemonitoring.

## 7.4. Shift in patient experience: stress and incoordination

Most human activities are related to the passage of time, from both a chronological and psychological point of view. However, the *measurement of time*, as the philosopher Fernando Savater outlines, reminds us of the *multiplicity of time*.

> *Measuring time is more or less equivalent to determining the duration of changes that affect us, in our activities, and which affect the world in which we live. But as such changes can be of many types, such as the measures that we apply responding to very different criteria, it is impossible in reality to speak of a single "time": we must resign ourselves to the fact that there are several "times", according to the observed changes and the means of measurement employed. And also*

*according to the social emergency which prompts us to control some of the changes above all others* [SAV 00, p. 271].

The subjective perception of time that transpires behind the perceived duration, synchrony, the pace of life and our temporal perspectives are intertwined. Joseph McGrath and Franzsika Tschan, researchers in social psychology, however distinguish four processes in the temporal dimension of everyday life: *time use*, *pace of life*, *time perception* and *time orientation*. The use of time refers to the distribution of the person's time in the course of a day: breakfast, work, leisure time, travel time and personal care. The pace of life is the speed at which the person performs daily tasks. The perception of the time dimension which particularly interests us in this chapter is the way that people measure the passage of time (the duration of certain time intervals, for example). Orientation in time refers to the way in which people compare the present to the future and the past – i.e. regrets, hopes and fears (see Box 7.3).

| **Time use** | **Pace of life** | **Time perception** | **Time orientation** |
|---|---|---|---|
| Time distribution on a daily basis (work, meals, recreation, travel) | Execution speed of daily tasks | Psychological experience of passing time | Self-positioning compared to time (present, past and future) |
| Chronos | Aiôn (user experience) | | Kairos |
| User interfaces | | | |

**Box 7.3.** *The four dimensions of the subjective perception of time according to McGrath and Tschan, adapted to telecare*[13]

Individuals' perception and orientation of time are negotiated in a different way in the context of telemonitoring, for two main reasons: on the one hand, biological mechanisms of temporal perception, which have just been described, are the object of a singular inflection (see analysis below); on the other hand, multiple temporalities specific to the telemonitoring system "overflow" by cutting across the general time frame of individuals. It is therefore appropriate to refine the generic design proposed by McGrath and Tschan by taking into account four constituent areas of the temporality of telemonitoring: (1) the body and disease, (2) objects and communicating systems, (3) care and treatments and (4) the players involved: caregivers and familial or social carers (see Figure 7.2).

---

13 [MCG 04]. See also Chapter 5, section 5.6.

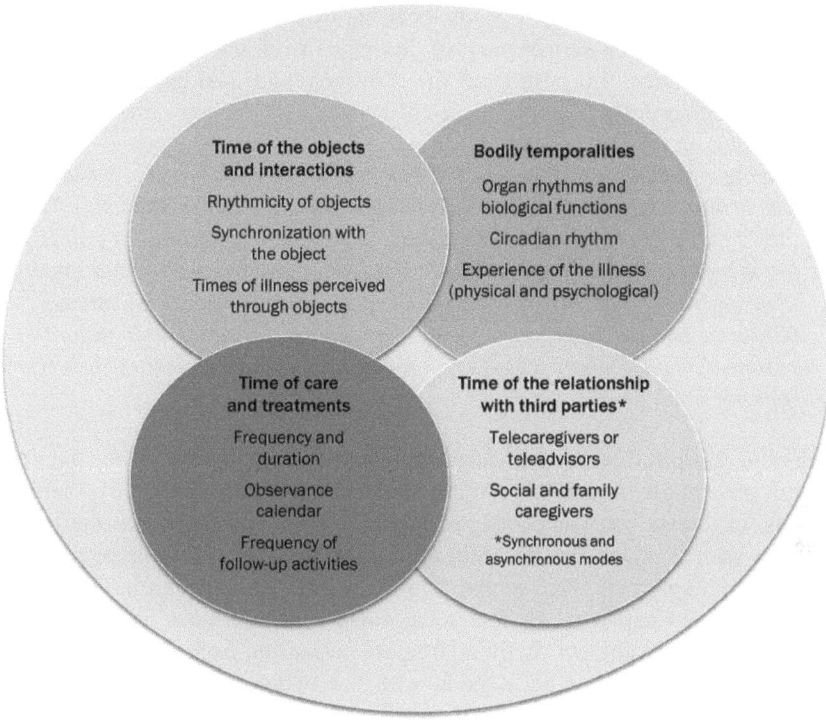


**Figure 7.2.** *Temporal plan of the telemonitored person at home*

Figure 7.2 presents a typology of the temporalities of the telemonitored person with regard to the biological foundations of biotemporality that have been previously outlined. "The psychological experiences of time passing", as described by McGrath and Tschan, in effect transforms itself into an experience of multiple temporalities conditioned by external factors, such as the use of communicating objects, that enter into the care equation exercising in return a certain influence on *the use of time and pace of life* of the telemonitored person as well as on their psychological experience and, as we will see, their *orientation in time*. In practice, the coordination of the person's relationships, at home or remote, with caregivers and/or social and familial caregivers; care and treatment; and the use of the communicating objects, whose purpose is to identify more closely the rhythms of the organs and biological functions of the person, can be seen as a synchronization process that imposes time constraints. Such coordination may influence the *level of stress* of the telemonitored person and therefore disrupt their circadian rhythm in the same way as day–night cycles, alternating rest–activity or meal times, as discussed previously.

Awareness at this level is already bearing fruit in the field of personalized coaching, which targets consistency of individuals in the management of their multiple temporalities, in particular the Chronos and Kairos time, mentioned previously (see Box 7.3).

> *During coaching, the work of the coach aims to make consistent and synchronize the Chronos and Kairos time for the trainee. We can cite the example of professional retraining. A person may feel that it's time for them to change. The coach will then bring out deeper motivations (Kairos time), and will work with them to pave temporal milestones (Chronos time) to allow for change. By aligning these two dimensions of time, Chronos and Kairos, experience shows that the person feels a sense of inner peace*[14].

The time constraint of their disease and telecare is a daily reality for the patient. Temporal perception of the telemonitored person may be particularly subject to conditions of stress and anxiety generated by the gap between the physical experience, such as measured by objects, and the psychological experience, or the person's subjectively perceived temporality[15]. In fact, the patient's biological clock undergoes constant and more or less severe disorders which are phase shifts from so-called "normal" biological rhythms. Remote measuring from this phenomenon of temporal discontinuity may be at odds with the manner in which the patient is intuitively considering their illness and how their long-term disease is represented to them. In other words, the digital representations of biological rhythms of chronic patients do not necessarily coincide with the intuitive representations of the disease and its temporality. This *inconsistency* is the trace of a realignment of feeling and the patient's emotional state on the intimate experience of their illness through the telemonitoring protocol. However, to better understand the impact of the affects on the patient's subjective time, it is essential to consider first the way in which emotions modulate the sense of passing time.

## 7.5. Affects as regulators

The influence of affects on the patient's temporal perception – especially on their psychological experience of illness and their self-positioning compared to time

14 http://www.3hcoaching.com/connaissance/les-3-types-de-temps-chronos-kairos-et-aion/.

15 Subjective temporality of the patient finds its ontological basis in the fact that the psyche interacts constantly with the body; this phenomenon can be explained biologically by relationships between neuropsychiatric structures and demonstrated physiological functions, following the discovery of the brain peptide hormones and neurotransmitters by Roger Guillemin, Andrew Schally and Rosalyn Yalow (winners of the Nobel Prize for medicine in 1977).

scales – is related to the *degree of physiological activation*. Physiological activation, or physical response to environmental stimuli, is sometimes called "adrenaline syndrome" because increased secretion of adrenaline and noradrenaline plays an essential role. The advanced hypothesis here is that the physiological activation triggered by environmental stimuli, especially mediation imposed with some communicating objects, may increase (in the same way as the changes in light and temperature referred to above) the intensity of the affects of the regulating patient, through the *interoception* (see Box 7.4), their physical experience of the illness and their telemonitoring experience.

As the consciousness of the internal state of one's body, interoception tells us about emotions, pain, thirst, hunger and body temperature. The affects find their biological basis in bodily (interoceptive) signals originating from the viscera affecting the limbic brain, and this action in turn modulates the individual's temporal perception. Philosopher Claude Debru thus retraces the steps that allow emotions to modulate our sense of time passing: "For various authors [...] this would be interoceptive sensory messages from the viscera involved in the organic reaction to the aggressive stimulus which (in humans) act on the insula, a cerebral structure belonging to the limbic brain [the insula is located in the depth between the temporal lobe and the frontal lobe]. This would be the essential area inclusive to (interoceptive) bodily signals, i.e. the area of conscious perception of visceral states, which would produce what authors have called 'emotional times'. However, experiments have shown links between temporal estimates and respiratory or cardiac rhythm" [BUS 11, p. 243]. Several theories of emotion confirm the link between the interoceptive process and emotions, including those of William James and Antonio Damasio[16]. Olga Pollatos highlights the interest of the dual heritage of James and Damasio: "William James postulated that viscero-afferent feedback is closely linked to emotional experience, stating that 'bodily changes follow directly the perception of the exciting fact, and that our feelings of the same changes as they occur IS the emotion' [from William James, "What is an emotion?", Mind, vol. 9, pp. 188–205, 1884] [...] An example for a recent psychological theory incorporating the feedback from the peripheral nervous system (somatosensory and visceral) is the somatic marker hypothesis by Damasio and colleagues [...] The authors speak of the obligatory body-relatedness of feeling: 'the body is the main stage for emotions, either directly or via its representations in somatosensory structures of the brain'" [POL 05].

**Box 7.4.** *Links between interoception and emotions*

The intensification of affects under the effect of environmental stimuli, well known in psychophysiology (see Box 7.5), can have an impact on the physical experience of the illness by inducing a more or less elevated stress in the patient,

16 See, in particular, [JAM 84], [DAM 94] and [CRA 04].

according to the degree of compulsion felt opposite to the imperative synchronization with their communicating objects and home telemonitoring system.

> Psychologist Jacques Van Rialer recalls the psychophysiological bases of this phenomenon: "The flow of our cognitions and our affects depends, among other things, on the state of our autonomic nervous system. This part of the nervous system automatically regulates the functioning of endocrine glands, the heart, blood vessels, breathing, the digestive system and the dilation of the pupils. It is composed of the orthosympathetic (or 'sympathetic') system and the parasympathetic system. The current meaning of the word 'sympathetic' in psychophysiology, corresponds to its etymology (*sun* – *pathos*): 'in connection with the affects'. In fact, all 'active' emotions stimulate the orthosympathetic system and produce a physiological activation. It is particularly intense in emotions that lead to vigorous actions: fear, anger, sexual arousal, amorous shock. It is less strong but still present in joy, enthusiasm, hope, disappointment [...]. Physiological activation increases the intensity of emotions. If we receive an injection of adrenaline in a laboratory situation, while we are quite calm, we will probably feel the physical symptoms (cardiac acceleration, feeling of warmth, slight trembling, etc.), but not a true emotion. However, if we perceive a threat at that moment, our fear will be significantly more intense than if we had not received this injection or if we had taken a sedative. If our sympathetic system is active and, on the other hand, we are irritated, we will more easily experience anger. If we see an amusing movie with a good dose of adrenalin in our blood, we'll laugh even more. If we see it having taken a tranquilizer, we'll laugh much less" [VAN 14, p. 226].

**Box 7.5.** *Description of the intensification phenomenon of the affects under the effect of environmental stimuli*

On the other hand, the intensity of physiological activation is intimately linked to the valence of emotional states, in other words to the inherently pleasant or unpleasant quality of a stimulus or a given situation (see Definition 7.2).

> DEFINITION 7.2.– In psychology, the term "valence" is used to designate the intrinsically pleasant or unpleasant quality of a stimulus or situation. In the context of the psychology of emotions, the concept of valence also refers to the pleasing character of certain emotional states (such as joy or contemplation of beauty) to which a *positive* valence is attributed as opposed to *negative* valence emotions (such as fear or sadness) usually associated with discontent, inconvenience or suffering[17]. Added to that are the stock *value* and flow *valence*

17 Source: https://en.wikipedia.org/wiki/Valence_(psychology).

used, for example, to analyze the number of patients (*stock*) with the flu (*flow*). The valence of flows, in particular, is one of the constituent elements of *transduction* according to Simondon (see Definition 7.3).

Therefore, depending on the nature of the emotional, positive or negative valence, the person's physiological activation will be more or less intense. Furthermore, some recent studies show that temporal perception is, on the one hand, faster and *governed by emotions* in a context of high physiological activation, and, on the other hand, it is both slower and *governed by attention span* in a context of low physiological activation. These conclusions come from the work, among other contemporary authors, of Alessandro Angrilli and the psychophysiology research group at the University of Padova:

> *In short, our results provide evidence of an effect on the temporal estimation induced by the manipulation of two primary components of emotions: emotional valence and the level of stimulation. The interaction between these two factors is in line with the coexistence of two different mechanisms for assessment of perceived time. The first mechanism is faster, activated by emotions and goes off in situations of high stimulation. The second mechanism is slower, activated by attention and occurs in situations of low stimulation. Although the low stimulation mechanism is fully taken into account by the temporal perception models based on attention, the interpretation of results through the psychobiological model is more probative. According to these models, temporal perception seems to be more strongly linked to the trend of action* [ANG 97, p. 980].

Thus, in the telemonitoring framework, the higher the degree of the patient's physiological activation is (due to their use of communicating objects, interactions programmed with a telepractitioner or alerts on their state of health) the more the intensity of their emotions increases and modulates according to their temporal perception of the disease and monitoring protocol. However, it seems difficult to assess, with precision, the degree of positive or negative valence of the emotional states of the telepatient resulting from the physiological activation triggered by these phenomena. No study makes note of the induced stress of the telemonitored patient by physiological activation, even if the results of the qualitative survey of telepatients, for the purposes of this book, seem to confirm this hypothesis[18].

---

18 See Chapter 6 for an analysis of the results of the qualitative study on the experience of chronic patients in a situation of home telemonitoring.

It should be noted that the valence of emotional states caused by telemonitoring can without a doubt be positively modulated by the therapeutic education of the patient or the progressive mastery of communicating objects. On the other hand, the valence of emotional states, which depends on the intensity of the physiological activation of the telemonitored person, does not necessarily constitute an obstacle to the development of their psychological experience of the disease. From a psycho-temporal point of view, the patient's adaptation to their telemedical environment can indeed bring a *desire* along with it, meaning an "active" emotion in the psychophysiological sense of the term[19], to project themselves into the future, beyond the limits of the established protocol. The epistemological scope of this desire can be measured against the psychological timescales.

This anticipation approach is part of a long history of intentionality, of which Alain Berthoz traces the outline:

> *In the past, anticipation has been at the center of the reflections of thinkers. [...] It appears in all philosophical literature on intentionality. We will reflect, of course, on the 'vital impetus' of Bergson, this arrow directed toward the future. But already, the Kantian a priori, i.e. the organizational framework of possible empirical knowledge, functions on the mode of anticipation: in this case, the "anticipations of perception" are the anticipation of a continuous gradation of the sensory matter of all phenomena. And what arouses the interest of the researcher in cognitive science for the description by Husserl of the conscienceness of time, is that he founded the constitution of the identity of objects of perception through their successive presentation on the game of retention of the just past and of protention toward an imminent future* [BER 15a, pp. 12–13].

## 7.6. The future in sight

| |
|---|
| QUESTION 7.2.– How can anticipation, as a cognitive act nurtured by affects, impose itself in the temporal consciousness of the telemonitored person? |

First, note that there is knowledge of anticipation that is relatively autonomous with respect to forecasts made as part of a medical telemonitoring protocol. This includes, in particular, self-intuitive knowledge, acquired by the chronic patient on a daily basis, influencing their condition, perception and future. Recognizing the

19 For psychophysiology, active emotions (fear, anger, desire) cause strong actions.

symptoms, identifying stressors and anticipating the effects: such intimate "know-how" is transmuting into personal, non-quantifiable knowledge of chronic disease. The uniqueness of this *subjective* knowledge, based on the affects, is due to their potential for activation of *retention* (reactivation of past situations) and *protention* (anticipation of events) of the person and their anticipation desire. In other words, the links that the person is able to establish with the future are induced by retention and protention mechanisms by which the individual, as neurophysiologist Alain Berthoz highlights, inspired by Edmund Husserl, perceives together their environment (and their illness) and the outside world, projecting their own desires and intentions:

> *This protention mechanism, a term introduced by Husserl, coupled with that of retention would allow the individual to treat their present environment by continuously combining past and future phenomena. The experimental results suggest that the perception would be based on internalization of a part of the world and would thus shape external reality, projecting their preconceptions, their desires, their intentions* [BER 15, p. 226].

This *creative intuition* of the world, which is built from the present time, around the categories of the past and the future, is completed by the *proactive intuition* of the person (ability to act, to represent the future) that suggests the prospect of a *cathartic* future, i.e. a future conducive to a remediation of their condition, coming to take over a present issue and a *pre- or post-symptomatic* past, and this, for one major reason: the coordination capacity with what can happen (the future) allows for an adaptation of the individual to their environment. Anticipatory intuition thus raises a natural impulse of the individual, close to the instinct of survival, which leads them to take the future in sight. The prevalence of the future in the temporal consciousness of the person in the context of telemonitoring is also due to the ability of the communicating objects to evoke a strong emotional response, which echoes the ability of the individual to perceive their environment through senses and to influence their cognitive ability, recalling the stimulating capacity of environmental factors described by psychologist Jacques Van Rialer:

> *External events do not identify us mechanically. They "inspire" us based on other variables, at the forefront of which is the cognitive dimension, specifically our way of perceiving, remembering and interpreting* [VAN 14, p. 26].

In addition, recent studies in cognitive science [PEZ 08, HES 02] show that the representation of the future, and how we coordinate our actions in relation to this, leads us to engage with the environment on a *simulation* mode. This is only possible, however, through internal predictive models that allow us to predict the consequences of our actions as well as the dynamics of the environment. According to neurophysiologist Giovanni Pezzulo, these predictive models and proactive mechanisms of human behavior that underlie them are necessary to explain the diversity of mental phenomena that produce a behavior that is not immediately motivated by the "here-and-now" in the environment, but by the "to-come" [PEZ 08, p. 181]. Anticipation allows us to generate distant goals: proactive representations that select and guide the action. To purposely coordinate many of their actions in time, the cognitive agent must be able to conceive and represent (implicitly or explicitly) their future, including their objectives, and maintain their predictions and most accurate possible representations. The categories of the present and past are not incidental. As philosopher Claude Debru highlights, "to mention the fact that the subject is sensitive, [...] in their 'protention' toward the future, to their environmental framework and atmosphere, inevitably raises the issue of the role of memory. Perhaps a working memory, but more likely a reference memory, referencing events or situations that had a continuation. Thus one can understand how memory of the past comes here to help the prediction of the future" [BUS 11, p. 137]. Nonetheless, the prevalence of the future in the temporal consciousness of the person promotes the development of an exit scenario symbolic of this *instantial present* (that of the chronic disease experienced daily) and the perspective of a layout of a *referential* history. Protention may allow the individual to activate proactive mechanisms that will provide them the opportunity to achieve a path of wellness by simulating it mentally[20]. In these conditions, anticipation as a cognitive act forms the basis of the *mental journey* of the person toward a new self-experience.

## 7.7. Anticipation as action project

How can the anticipation, as *movement of thought*, allow the patient to develop a *self* project independent of their involvement in a care protocol?

Anticipation is a fundamental activity of the living. Achieving a goal requires prior determination of the said goal, actions to accomplish it and the path to take. For the individual, a "goal" is any object likely to contribute to their well-being, and, in the case of the (chronic) patient, to protect their health. For psychology, anticipation is not reducible to a single cognitive act: it is a movement of thought that involves the identity and the integrity of the individual to project in the future. In the context of home telemonitoring, anticipation allows us to operate a registry change: from preventive management of the body generating a temporal-existential break–that of the patient who thinks of themself without any future because they

20 For Husserl, as Claude Debru highlights, protention means "consciousness which, in its own acts, surpasses the present into the future that it anticipates" [BUS 11, p. 65].

are already predetermined by the chronic disease ("entering into chronic disease", explains the philosopher Philippe Barrier, "is a brutal understanding of one's proximity to death") [BAR 09] – to the subject's assumption of their own destiny. As such, anticipation can mean to register and advance in a movement of individuation or subjectivation. In short, according to psychologist Marcela Gargiulo, "If prediction is the responsibility of the medical act, establishes itself and dissolves uncertainty, anticipation is a quest for the subject to appropriate knowledge that is revealed. The threat of a future disease reaffirms that time is counted and can lead to future misleading representations. Everything happens as if the thought of the loss of the future constituted an absolute knowledge, which was already integrated" [GAR 14, p. 189]. Therefore, protention is not only, and ordinarily, the anticipation of *a coming* but the anticipation of a cathartic event: the realization of a *self action project* through which the possibility conditions of *self continuity* can be created, beyond the limits of chronic disease.

At this stage of the study, the psychophysiological examination of biological rhythms, of multiple temporalities in telemonitoring, of the specificity of performative affects, of the patient's temporal perception as well as the analysis of their proactive mechanisms and the psychological timescales – designated here as "anatomy of anticipation" – allow us to better understand the direction toward the future of the person, which is a phenomenon specific to the conscienceness, as assumed by the elaboration of a self project. If the analysis of this *mental journey* by proactive mechanisms confirms that *the time for the telemonitored person is a subjective property to build possible experience*, it raises the essential question of the role that the communicating objects and, more broadly, the telemonitoring system are expected to play in the subjectivation of *their future*.

## 7.8. *"Affordance"* as an anticipation vector

One of the first psychologists to integrate the role of action possibilities in the perceptual process was James Gibson [GIB 79]. He applied an interdependence between perception and action, including the direct input of the action in the perceptual process. Gibson coined the term "affordance" to refer to the ability of an organism to perceive action possibilities offered by its environment. According to Gibson, these affordances would therefore depend on the properties of both the environment and the organism. The affordance concept therefore seems relevant to identify the role of communicating objects here in the subjectivation process of the patient.

The use of communicating objects to ensure remote care for the patient is indeed an accompaniment, or a facilitation of individuation with the chronic illness, but it is also a possible *shifting mode* in the subjectivation of the becoming of the person.

The nature of this shift is mainly due to the possibilities offered by communicating objects that can be seen by the telemonitored person beyond their "functionality". This potentiality is first that of *transductivity* (see Definition 7.3), which is the ability of objects to bring about specific structures (the autobiographical consciousness) by which activation of *potentialities of being* occurs (also called "vicarious affordances")[21]. These "self-affordances" can be linked to the past of the individual (such as the opportunities missed, for example) or associated with their future (anticipating the benefit of future opportunities). These self-related potentialities, seeded in the unconscious and constitutive of a story about oneself (which could have been or could be conceived by the person), also capable of acting on representations of a future self, are brought to consciousness by the communicating objects, or more specifically by the affordance of transductivity (see also the definition of transduction according to Simondon, Definiton 7.1).

DEFINITION 7.3.– For Gilbert Simondon, the principle of transductivity denotes not only the advent, but also the individuation "without end", of specific structures: "[transduction] is a physical, biological, mental and social operation whereby activity spreads gradually within a domain. This propagation is based on a structuring of the area operated from place to place, where each region of the established structure is used for the following construction principle region, and so a modification extends gradually at the same time as this structuring operation. A crystal which grows and extends from a tiny seed in all directions in its mother liquid provides the simplest picture of the transductive operation. Each molecular layer that is already established serves as a structuring base of the layer forming: the result is a broadening reticular structure. The operation is an individuation in progress" [SIM 05, pp. 32–33].

By daily use of communicating objects, the patient's anticipatory intuition inscribed in the unconscious can thus change: it is more than a mere intuition from the body, but an intuition *inflected* by temporal disease management that takes place through communicating objects. In a recent book, Claude Debru and Pierre Buser break down the psychological mechanisms of anticipation, "which could take various forms, waiting for a stimulus or event, waiting with positive emotional engagement, with the hope and pleasure of a reward, or negative, with the dreaded fear of undesirable punishment, preparation of a motor act", and show that "these markings of anticipatory structures in the brain were sometimes neural, therefore electrophysiological, and sometimes image makers, and demonstrated that, in certain cerebral structures, neural circuits exist that witness this proactive property and this

21 Conceptualized by philosopher Aaron Sloman in the wake of the work of James Gibson (see [SLO 06]).

perception of the existence of a future"[22]. It must be emphasized here that the patient's proactive insight is a hybrid mental process that is located at the crossroads of "the Bayesian concept of conditional probability, i.e. a more or less high probability that something might occur in light of what marks the moment [probability that an alert is sent to the telepatient according to their state of health at a given time] and the Popperian concept of propension, i.e. the expectation of what is likely to happen, which is permanently present in our mind [probability of a complication with the disease]" [BUS 11, p. 279]. Between probabilities and propensity, anticipatory intuition that guides subjectivity seems to have the ability to start a future perspective for the person, beyond the limits of the present of their illness, taking support from the potentialities related to oneself ("vicarious affordances") "activated" by the transductivity affordance of the objects, mentioned earlier.

The role of objects in the subjectivation process of the future instituted by the patient can also be understood through the prism of the relationship between the objects' affordances and the person's episodic memory of the future[23], as defined by philosopher Denis Perrin:

> *In psychology we define episodic thinking of the future (ETF) as the ability to mentally experience, in advance, singular personal events that might occur in the future. Just like episodic memory, it allows us to perform experiments by substitution, form mental representations including an empirical phenomenon in which the subject that forms the representation is at the origin of the represented experiment. In the same way still, it puts into play the specific type of consciousness that is autonoetic consciousness, meaning the anticipated scene is experienced by the subject of anticipation as an extension (or projection) of their present self and consciousness of that extension towards the future* [PER 15, p. 195].

DEFINITION 7.4.– We generally attribute the authorship of the concept of "episodic memory", of which Denis Perrin gives the following definition, to cognition neuroscientist Endel Tulving: "[Episodic memory] is defined as [...] the ability of the subject to refer to an event that they have personally experienced and to mentally relive this experience. It thus differs from semantic memory – or the ability to know facts, not necessarily autobiographical, in the form of representations that do not seek mental reactivation of any past

22 [BUS 11, p. 278]; moreover, Pierre Buser and Claude Debru illustrate how the forms and markings of anticipatory intuition induce awareness of the probability of the moment that follows.
23 For a recent formulation of this concept, see, in particular, [SZP 10].

experience – in this it puts into play a form of specific awareness that accompanies episodic retrieval, namely "autonoetic consciousness": during episodic recall, I am conscious of mentally reliving an experience where *I was* the subject".

Episodic memory *of the future* which allows us to experience in advance personal events that are likely to happen in the future, just like episodic memory [PER 15, p. 193], can be activated through affordance of "projectivity" of communicating objects, because of the potential of these objects to awaken the consciousness of the patient's future by projecting themselves into the future through temporal knowledge about their disease. By this affordance, the person can be provoked to construct a sense of becoming, beyond the present boundaries of the disease. From a phenomenological point of view, the experience of the affordance of projectivity is one of the conditions for constructing a future self that "deports" the telemonitored patient out of an obligation to act on themselves under a telemonitoring protocol. So, the "future self" is not, or not only, an individual who "projects" themselves in the future, but a person who *commits their being* in the affirmation of a desire to escape the spatiotemporal limits enforced by the disease and, consequently, to symbolically distance themselves from the self-monitoring protocol imposed on a daily basis for preventive purposes.

This commitment of the being is not the trace of a refusal of the telemonitoring system and its technical constraints, more or less negotiated, or more or less accepted. It is not the sign of a subjectivity delivering an internal fight against the disease using interposed objects. This commitment is mainly a reflection of the patient's ability to define a self-project to reclaim the objects as "affordances" allowing them to re-appropriate their *bodily memory* and *autobiographical memory* in self-projection toward the future. The use of the communicating object is thus put into perspective: it does not only assist the patient with the quantitative measuring of their illness but it helps them project, or more exactly realize, their full protention potential by exteriozing it, and thus create the possibility conditions of their subjectivation of the future in the context of a self-project. The particularity of this project, as we will see, is for the telemonitored person to define *themselves* as a project: it is not for them to only lead a fight for health, but to find *self-meaning* in a *self-project* logged in their retention. It concerns self-identifying as a life project: capitalizing on one's situated knowledge and enhancing the sense of their "experience" as a patient. To better identify the epistemological scope of this "patient–project" concept, we propose to ask ourselves beforehand about the patient's status as a "subject to become" in the telemonitoring process, which is, once again, asking about the possibility conditions of an inflection of the therapeutic view on the patient as a person and a reclassification of the vision of care as human solidarity.

## 7.9. The patient as *poiesis* and *self-project*

*This is the only force of the past coming from the future: somehow I live or I appreciate my past, I can only do that in light of a self-project concerning the future.*

Jean-Paul Sartre [SAR 01, p. 544]

DEFINITION 7.5.– Maieutics (from the ancient Greek μαιευτική, by analogy with the character in the Greek mythology, Maia, who watched over childbirth) is a technique that involves asking a person to make them express (deliver) knowledge. Maieutics consists of producing the essence of their knowledge. It is intended to express a knowledge hidden in oneself.[24]

The use of connected devices in the context of telecare leads the patient to wonder about their "function" within the system, as well as their status as a person. The relationship between the patient and the doctor in this induced context leads to, as has been mentioned, a co-construction and sharing of knowledge that enlightens the person about their state of health and duty. This maieutics approach, induced by interposed objects and empathic listening from the doctor, establishes a practice of care that draws attention to the patient's status. It concerns each actor in telecare in order to "produce" knowledge about one another through dialog. This "relational maieutics" informally builds on the fringes of "cognitive maieutics" set up in the 1980s, as part of the development of expert systems. Cognitive maieutics concerned producing knowledge in order to define rules allowing us to operate an information system and share data in a network. The patient, as an actor, was not taken into account directly. Without a doubt the limit of the artificial intelligence systems today, the natural extension of expert systems based on cognitive maieutics, is that they do not recognize the "relational maieutics" mentioned above: they thus call for a recharacterization of the human–machine relationship, most notably in the context of telemonitoring chronic diseases.

Reconciling medical technicality with the humanity of care (relational dimension) implies recognizing that the patient is not only a somatopsychic being, but they are in *perpetual tension* between the soma and psyche described by Philippe Barrier:

> *Whatever the will of objectification of the patient by the doctor, and as important as the progress of technical and objective methods of observation and analysis might be, the patient can never be reduced to their purely material objectivity. But before even possibly being an*

24 Source: https://en.wikipedia.org/wiki/Socratic_method.

> *ethical requirement [...] it is first a medical reality: the disease affects a subject, and not a simple organized living matter, a somatopsychic being, an chiasmus that is irreducible to one of their components, a sort of bipolar being whose soma and psyche would be the extremities. There is not only always an echo but a resonance, meaning a return effect, of the alteration of one of the poles on the other [...]. And it is the clinic as an empathetic observer and listener for the patient that is the appropriate medical response to the call of this complex being* [BAR 10a].

The perpetual tension between the soma and psyche experienced by the patient is thus responsible for the displacement of the way in which the person looks at themselves. This look at the evolutionary resonance between the disease and their perception is, however, so sharp that it may be jointly considered a technological mediation. This mediation can be at the origin of a dephasing of the individual with respect to themselves as demonstrated by the analysis in Chapter 5: "The person may fall out of step with respect to themselves and their illness under the influence of the telemonitoring dynamics. On the one hand, this phase may induce a potential withdrawal and the possibility of a split, for the patient to develop a *'double' awareness of their illness*, in *and of* themselves, but also leads to antagonism between individual perceptions *and* technical disease"[25]. In fact, the soma and psyche doubly under tension from patient experiences and their interactions with one or more connected objects at home. It is through these medical devices that the hierarchy of (digital) knowledge of the disease coming to enlighten the self-awareness of the patient is constructed and organized. With the "epistemological machines" therefore looming, by their presence and their function of elucidation of the disease and its symptoms, on the perception of the individual, the connected objects also offer a view of the patient as *poiesis*, as an *anthropotechnic construction*, in the image of the person described by the philosopher Stéphane Vial: "The being is a *poiesis,* meaning an anthropotechnic construction [...] To exist is the result of a production. And technology, combined with other factors, takes an active part, to avoid saying major. Man is no longer an essence or a separated substance, but a process made, constantly to do" [VIA 13, p. 26].

As the sick person is not only a simple organized living matter, the patient, as *poeisis*, is not only a singular process but a *subject to become* whose values, beliefs, habits, representations and agencies accompany their experience of subjectivation of the disease. This approach of the patient as a process and biographical being implies the reaffirmation, in the wake of Georges Canguilhem, of the humanist and relational

25 See Chapter 5.

vocation of medicine, of which telemedicine renews the stakes, against its objectivist and reductionist temptations:

> *It is not [...] an objective method that pathologically qualifies a particular biological phenomenon. It is always the relationship to the sick individual, through the clinic, that justifies the pathological qualification. While acknowledging the importance of objective methods of observation and analysis in pathology, it doesn't seem possible to speak, in any logical correction, of "objective pathology" [...] The [objective] intention of the pathologist is not that their subject is a matter emptied of subjectivity* [CAN 10, pp. 156–157].

While the chronic patient, like any individual, is a *constructed process*, an anthropotechnic figure which cannot be reduced to its purely material objectivity, they are no less a sick person whom medicine must no longer attain as an object but as the care *subject*; i.e. in their experience of subjectivation of the disease and at the gathering of symptoms, which are both key to capitalizing on the pathology (for the doctor), signs to build the experience of illness (for the patient) and the “reality of a personal history, just as biological as emotional, cultural and social” [BAR 10a, p. 161] which is likely to offer new perspectives to the doctor:

> *The doctor who really wants to cure cannot interpret alone and only objectively (in a purely scientific objectification) the pathological phenomenon. This is therefore a topic of an opening. It must therefore associate the patient to a discovery that is also (and above all) their own* [BAR 10a, p. 161].

However, the openness of the doctor to the telepatient in terms of biographical being does not imply for the practitioner to deviate from the symptoms by polarizing themselves on their subjectivation, or focusing on patient experience at the expense of an “objective pathology”, but to maintain tension between these two extremes by identifying, as close as possible, what seems to be able to be constructed through experience and the patient's therapeutic course: a subjective conscienceness of the disease and its temporality, techno-medically *inflected* and forward-looking, in other words, the advent of the patient-project.

## 7.10. The project as a balancing dynamic

The “patient-project” concept is open and dynamic and allows focus on the behavioral level, and not only on the clinical or technical aspects of the disease and the course of care. The patient-project concept refers to the person who, having reached a level of normative consciousness of the disease and its temporality, tries to take the best advantage of the connected objects to sketch a horizon of expectations for themselves. Overshadowed by the chronic nature of the disease, the temporal horizon of the telemonitored person is actually *potentially* open to the opportunity to gain new knowledge and new know-how, including the ability to discern the nature of the risks brought by the disease and the ability to remedy them by anticipation.

The psychology of anticipation shows that the patient projects themselves into the future considering the nature of the risks – "subjective" and "objective" – of which their affection is a carrier. These two types of risks, however, do not necessarily coincide. If the "subjective" risk (or experience), as Marcela Gargiulo notes in *La psychologie de l'anticipation*, "is the individual and original representation that each individual builds on their specific risk" [GAR 14, p. 184], the fact remains that they "do not always correspond, point by point, to the objective risk of 50%. It is determined by other elements than those of medical knowledge" [GAR 14, p. 184]. What Marcela Gargiulo reaffirms here is the principle whereby the patient builds a vision of their own future according to their own perception of risk, from personal knowledge of the disease, which was discussed above.

The objective risk, in terms of which the possibility of complications are evaluated, or conversely, the improvement of the state of health, is however not severable from the subjective risk. In fact, connected objects support, because of their medico-preventive functionality, the coexistence of these two types of risks within the protocol. As soon as the patient connects to the telemonitoring network, any quantifiable measure of their state of health can influence their subjective perception of risk, and therefore their perception of their future. The ability of connected objects to anticipate the future of the patient on the basis of an objective risk therefore allows us to modulate the meaning that the patient places on the risk they perceive based on personal criteria. If the objective risk, resulting from the quantification of the body by the objects, is often likened to a form of reductionism, it may however collectively constitute a former knowledge of anticipation in the face of the uncertainty of the (chronic) condition, thus re-characterizing the belief that the patient's future would only become a determined and closed time[26]. This is essential because finding *self-meaning* in a *self-project*, registered in health care protection, is a subjectivation or individuation process that depends on technical mediation, partially emancipatory through the affordances, as we have seen, of the anticipatory mechanisms of the patient.

---

26 Philippe Barrier confirms this "diagnosis" for diabetic patients: "Any projection in the future seems almost impossible. 'Tomorrow' is thus practically only the continuation of a final 'today'" [BAR 09, p. 44].

In short, an extension of technical preventivity in the human consciousness and the anticipatory ability of the person to determine their future, because of the shift in affordances, can contribute to the development of a *self-project*, which refers to the self-reflective ability of the patient to determine their own future, by acting on themselves. This project, as a *potentiality*, can emerge in confrontation with the disease, in the relationship to the connected objects and in the evaluations of the risks (carried by objects), and ultimately lead to the construction by the patient of their own temporal normativity. By establishing the norm of their own future, by anticipation, the patient is thus brought to recharacterize their current state of health and feeling toward their illness, in a context of progress, and to affirm their ability of subjectivation of the future, enabled by and for themselves, as well as to draw a dividing line between the opposing forces of (technical) perception and (psychological) feeling. The psycho-emotional process, in which we examined the biological foundations and self-project, potentially creates a *dynamic balance* between the human body and the machine, the object and the care subject, the objective and subjective time (of the disease and its future), the calculated and perceived risk, and (systemic) prevention and (psychophysiological) anticipation.

The realization of a self-project finally allows the recharacterization of the values of the *possible* and *probable* and their specific features, as philosopher Fernando Savater suggests here:

> *The possibility arises from freedom and imagination; the probability arises from the statistics and subjugation to the "objective conditions". The possibility is chosen and claims to force itself against determinism and the monotonous rerun from what is common and current, while the probability is imposed on us from the outside and only offers another landmark that desperate repetition embodies. The likely approach cannot create any genuine reform of the current order because the "crumbs" of possibility that it still retains are rendered sterile by the oppressive routine of the order of the given; when something manages to be probable again, it means that the new developments have disappeared, and its promises even start to fade. However, the possible still actively questions the need for necessity and disdains of closed speech based on observed realities that have been verified a thousand times, only observing another collective project already seen* [SAV 05, p. 341].

The self-project as a dynamic of balancing takes into account the probable, in other words the nature of the risks associated with chronic disease (calculated risk) without, however, granting prevalence. It includes the *probable*, such as a piece of data required for judgment that aspires to achieve the *possible* in light of all the parameters mentioned above (human–machine interaction, subjective and objective temporalities, prevention and anticipation, etc.). In brief, the project itself operates the probable and possible and merges on experience and situated knowledge.

## 7.11. The meaning of the joint entry of the temporal and emotional

The person who is telemonitored medically has, like any patient, valuable intuitive knowledge of the disease, prior to any medical conceptualization; intimate knowledge of their condition and of its temporality; and a strong ability of anticipation, biologically determined and technologically oriented. These underpin their power to act and to define for *themselves* a future project beyond the limits of their illness. Telemonitoring objects, by their connecting functionality and their affordances of transductivity and projectivity, can influence the patient's temporal perception, their sense of becoming and ability to advance and other potentialities related to self-esteem (*vicarious affordances*).

The study of the anatomy of anticipation, or the biological foundations that govern its destiny in the context of medical telemonitoring, in particular the factors of stress and emotions, shows the importance of retention mechanisms and the sequence by which the patient perceives their illness and environment, as well as their future by projecting their own desires and intentions. In that respect, what appears through this careful consideration is the psychophysiological origins of the subjectivation of the future that make up the movement of thought offering to the patient the possibility of achieving a path of well-being, by simulating it mentally. By illuminating the cognitive specificities of anticipation, the process of subjectivation or individuation, psychology and more generally cognitive sciences allow us to explain the passage from the temporo-existential breakdown caused by chronic disease to the assumption by the patient of their own destiny, and this opens up the possibility of considering the anticipation as an action project.

The meaning of this subjectivation process of the future, which can potentially be triggered by the patient in the context of monitoring, also appears through the study of the relationship between the affordance of objects and episodic memory, which allows the patient to experience in advance personal events that are likely to occur in the future. The study of this report shows how a telemonitored person may subjectively build a sense of becoming beyond the borders of the disease. In those circumstances, the representation of a "self-future" can no longer be simply that of a patient who is projected into the future but a person engaging their being in a mental voyage to a new self-experience.

By considering the possibility conditions for such a journey by the subjectivation of the future that takes place through proactive mechanisms of consciousness and affects, the telemonitored patient is able to highlight a *self-project* on the future. Acknowledging this phenomenon allows the hope to be maintained of a possible reconciliation between the medical technology and the humanity of telecare, and it also allows us to review, in the final analysis, the status of communicating devices as "chrono-technologies" from the perspective of ethical re-qualification of the care of the temporality *and* the emotional health of the telemonitored person.

# 8

# "Chrono-sensitivity": From Concept to Ethics

*Ethics is what the subject stands on, refusing to become an absolute individual, a closed domain of reality with detached singularity.*

Gilbert Simondon[1]

On the one hand, mediation provided by the objects is symbolic – these are sense, value and emotion carriers and make it possible, as discussed in Chapter 7, for the projection of the person in the future, in the sense of building a self-project beyond the contingency of treatment. Mediation provided by telecare devices is also concrete because it weighs on the various ranges of temporality and emotions of the telemonitored person on a daily basis. Coordination and communication instruments, in real-time, body temporality measuring devices (vital signs, sleep activity, heart rate, etc.), vivid memories of the body, in the sense of physical data storage capacity (integrated chip defibrillators, biosensors on the skin, electronic tablets and smartphones used for archiving and for the management of the care calendar); the communicating objects are "time machines" or what we might provisionally refer to as "chrono-technologies".

Communicating objects paired to care devices – or defibrillators that perform a double function (remote monitoring and cardiac activity support) – do not directly contribute to strengthening the links between the person and the medical team but they, nevertheless, facilitate the caregiver–care recipient relationship by saving time for stakeholders: accelerated reading and transfer of physiological data, simplified

1 [SIM 05]. See also, concerning the ethical relationship specific to care, Philippe Barrier, [WOR 15, p. 157].

care procedures, tightened coordination and communication. The techno-therapeutic optimization of protocols allows a refocus of telecare, if necessary, on its human dimension. In other words, the efficiency of the telecare system is not an end in itself, but rather the means by which carers and care recipients can make time for the relational quality of care.

In what ethical sense can we therefore grant a complementarity of temporal *and* relational care in a context of telemonitoring? What light can epistemological analysis shed on the temporal and relational organization that takes place through telecare, at the origin of the "chrono-sensitivity" concept? How can this concept shape *the ethics of telecare centered on the patient*[2]?

This chapter is intended to answer these questions and to explain the concept of chrono-sensitivity by borrowing from *care* ethics, in the wake of the work from American psychologist Carol Gilligan and, to a lesser extent, from the current ethics of personalism. Like the present study, these two trends have the particularity to register the *person* and their values at the heart of care ethics.

In this perspective, it is recommended to inform beforehand the temporal and relational aspects of telecare.

## 8.1. Telecare as "chrono-technology"

Objects used for the telecare of people with chronic illnesses are a subset of "new technologies" used today in the field of digital health, listed in Table 8.1.

| Categories | Objects | Industry actors | Pathologies |
|---|---|---|---|
| Implantable/integrated | Computerized pacemakers<br>RFID chips<br>Biosensors | SORIN<br>ORANGE HEALTHCARE<br>EQUIVITAL | Heart failure<br>Renal failure<br>Other pathologies |

2 As a reminder, the *patient* concept used here as part of the ethical reflection on telecare is defined in Chapter 2: "The meaning given to the person, at the center of a relational ethic of telecare, is thus threefold: it refers to the trying subject to, the individual co-responsible for care and to the being who reinvents themself (here under the influence of chronic illness). This definition of the person draws attention to the personal being who can no longer be ignored in the care relationship".

| | | | |
|---|---|---|---|
| Ingestible | Electronic pill | PROTEUS | All pathologies |
| Connectable | Electronic devices for health and welfare<br>(connected blood pressure monitor, telemonitored removable prostheses, etc.) | WITHINGS<br>iHEALTH<br>BODYTRACE | All pathologies |
| Portable | Health bracelet<br>(Bio) sensors<br>Smart watches | iHEALTH<br>WITHINGS<br>APPLE | Arrhythmia<br>Vital signs<br>All pathologies |
| Next Gen portable | "Connected" clothes | BIOSERENITY<br>CITIZEN SCIENCE | Epilepsy<br>All pathologies |
| Tattooable | Electronic tattoo*<br>[*in development] | UNIVERSITY OF AUSTIN | All pathologies |
| Activated | Computerized insulin pumps<br>Pumps for chemotherapy | MEDTRONIC<br>VOLUNTIS | Diabetes<br>Cancer |
| Interfaceable | Tablets<br>Smartphones<br>PCs/watches | MICROSOFT<br>SAMSUNG<br>APPLE | All pathologies |

**Table 8.1.** *Selection of smart medical objects grouped by categories and industrial actors*

The rapid proliferation and adoption of these new health technologies show that telecare is not simply a trend of telemedicine but an expanding "market". To its critics, telemedicine seems doomed to eclipse conventional medicine. To accept that view, however, would indicate how fast we forget the intrinsically human dimension of diagnosis: palpation, gaze, listening to the body. These actions, among others, will never be ignored during diagnosis or in care, even if they are delegated to the telepatient becoming co-caregiver. The other developed argument by opposers of telemedicine, including in the medical field, is about the humanity of care: the use of machines at home would depersonalize the caregiver–patient relationship. The depersonalization risk of the relationship, while feasible, may nevertheless be rejected if the conditions for the implementation of *responsible* telecare are met.

It seems that independent care is evolving today in the sense of a reinforcement of human assistance, each time possible, *in addition* to technological monitoring.

Psychological support and doctor–patient dialogue must, according to patients, be proposed, maintained and reinforced at a time when telemedicine is in the processes of becoming mass medicine. In short, it is vital to preserve the relational dimension of care, but also to pay more attention to the many facets of the temporality of the person and telecare which has been questioned thus far. As we have seen, care for the *patient-as-person* involves recognition of their physical and psychological experience of the chronic illness. Such care also relies on the volunteer work accomplished by caregivers who support them by providing strong psychological support.

Telecare devices (of the connected defibrillator type, or smartphones connected to a computerized insulin pump) and/or separate communicating objects (tablets or smartphones with companion health software) are dedicated to managing the various temporal aspects of the disease and to facilitate the monitoring of daily care at home. In this sense, these multiple objects are *chrono-technologies*: they are perceived as shaping the various temporal aspects of the disease, care, the patient and the doctor–patient relationship. Telecare chrono-technology allows stakeholders to *timely* identify, in particular, the biology of the disease and/or body condition: blood sugar, arrhythmias, weight before being restored during home hemodialysis, etc. Good *timing* and the accuracy of these statements depends, to a large extent, on the quality of care managed independently. Chrono-technologies allow, on the other hand, to ensure, for the better, optimized coordination of the monitoring process as well as the improved management of the time devoted to care: the insulin pump runs faster than the patient who pricks themselves, the connected defibrillator informs the medical team of the patient's heart activity in real time and the cycler optimizes the duration of independent home daily dialysis. Finally, chrono-care technologies, as we have seen, allow a harmonization of knowledge in real time: physical data can indeed be shared in a synchronous mode by all actors in the telecare relationship. In addition, optimization of the temporality of care positively impacts the quality of the caregiver–patient relationship. It remains to be seen what status can be granted to chrono-technology from the point of view of the philosophy of technology.

As a prevalent symbolic (Cassirer), ideological (Habermas) or metaphysical (Heidegger) form, the technique, in the "classic" philosophy of technology, is always a relationship with our habits, thoughts and know-how. As part of telecare, the philosophical question of the technique is that of a thought that emerges from the relationship of the individual with a material and medical culture whose objects are "agents" in the service of "actors". However, regardless of the point of view of philosophy concerning the technique, any mobilization of technical objects, and even more so of chrono-technical objects as their avatars, is a *production of its own*.

This production is at the origin of the problem of technological determinism from which the question of the relationship between the individual and technique is considered. This relationship in the context of telecare is temporal and relational. It is through an *empathetic* management of care within a chrono-technology system ensuring the objective measure of temporality of the disease that *self-realization* may be operated. The relationship between the individual and the technique does not register here in the opposition of subjective and objective knowledge about the disease, facing a technological determinism, but with the combination of these two types of knowledge because of telecare.

The relationship of the individual with the technique, in the context of telecare, from the point of view of the philosophy of technology and in accordance with the main thrust of this book, must be entered through the arrangement of *objective* temporality (that of the body and its measure) and *subjective* temporality of the person (experience) by interposed communicating objects, but also through the articulation of the *temporal* and *relational* dimensions of remote care that represents a major ethical issue for contemporary telemedicine. The mobilization of chrono-technology objects, associated with psychological support for people, through responsible practices of telecare, grants equal ethical value to these two dimensions (also see "chrono-sensitivity": a hybrid concept, in the following). The ethical recognition of temporal and relational values of telecare confer a status of *moral* agents to chrono-technology objects, or more precisely a status of *ethical agents*.

The concept of a moral agent means, in the sense proposed by philosopher Luciano Floridi, the entity that can, on the one hand, trigger a moral action and, on the other hand, cause *moral good* or *moral harm* [FLO 11]. Luciano Floridi sets himself apart from other contemporary theorists of moral agency such as Deborah Johnson [JOH 06] and Kenneth Himma [HIM 09], for whom agents said to be *artificial* (computers, software, electronic devices) cannot be moral agents since these have no freedom, consciousness or intentionality. Chrono-technology objects can, nonetheless, be seen as *normative* agents within the meaning given by philosopher James Moor [MOO 06]. According to Moor, an artificial agent (computer, computerized device) can in effect be described as a normative agent from the time it is possible to evaluate the task(s) that has/have been assigned in ethical terms: such is the case of telecare chrono-technology, such as the patient's autonomy. With this principle, Moor offers the following classification, known as the *four levels of ethical agents*:

– *ethical-impact agents*;

– *implicit-ethical agents*;

– *explicit-ethical agents*;

– *full-ethical agents*.

Care chrono-technologies are not only normative agents, within the Moorian sense of the term, in which it is possible to assess the performance level; these are also agents with an ethical impact, more or less explicit depending on the steps of care and the *timing* in which they are required to ensure monitoring in the context of telecare. Thus, a computerized insulin pump is *programmed* to trigger an injection at the right time and to limit the potential harm that might be caused to the patient by hypo- or hyperglycemia. The ethical function of the pump is to limit, in good time, physical harm that can be inflicted on the person with diabetes. In Moorian nomenclature, the insulin pump can be considered to be an *explicit* ethical agent. In other words, it is possible to evaluate its impact on the person in explicitly moral terms. The cycler and the defibrillator, by their techno-therapeutic features, belong in the same ethical agents category. On the other hand, the ethical impact of a smartphone or tablet type object, used in a context of telecare, is not explicitly measurable. These objects allow stakeholders to *coordinate* and to comply with the protocol standards, but they are not equipped with an integrated therapeutic function allowing them to engage an *explicitly* moral action. In Moorian classification, communicating objects can therefore be considered *implicit* ethical agents. However, the Moorian classification is only meaningful here, provided that it informs the ethical dimension of the *temporal and relational arrangement* that takes place through chrono-technology objects.

## 8.2. Telecare as a relational practice

The moral understanding of chrono-technological objects finds a special meaning in the context of telecare because it applies jointly to the temporal *and* relational dimensions of care. The latter concerns not only the relationship between the caregiver and the care recipient in the singular meeting practiced within the margins of telecare (monthly in the case of the monitoring of patients in independent dialysis at home, for example): it is also instantiated in the digital communication that connects the caregiver to the care recipient. So, the data transmitted by the defibrillator can result, depending on the alert level sent to the doctor, in the patient's moral support. The daily record transmitted to the nephrologist by the person following independent home hemodialysis or blood glucose levels passed by the diabetic person to the medical team can achieve the same result. In these three cases, chrono-technology objects have a similar mediating function that does not only raise communication: they make a moral action possible – *taking care of the person* – to which the caregiver is responsible. Alert messages, patient electronic communications, data reported by the nurses who provide care at home or remotely are usually completed through communicating objects and are the subject of a remediation, more or less formal, provided by the doctor. Through electronic links, care recipients can break their isolation and caregivers can identify, as close as possible, the evolution of the disease and the person's psyche. Through the use of these technologies, it is therefore possible to establish a level of trust between the actors and a more or less empathetic relationship that contributes to the well-being of the person. Implicit-ethical agents status of communicating objects, from which a personalization of electronic tracking can operate, is thus justified. What is also reaffirmed, through the moral agency of objects, is the personalizing goal of telecare – which designates therapeutic *actions* made by the patient *and* attention given to the *singularly tested individual* – in its double temporal and relational dimension.

## 8.3. Chrono-sensitivity: a hybrid concept

Understanding the *chrono-sensitive* nature of telecare involves recognizing the importance of the relational specificity of care, in its dual digital *and* human dimension. Identifying the chrono-sensitive nature of telecare also requires the recognition the *equal ethical* value of the relational dimension, thus defined, and the temporal dimension of the disease and the patient involved in telecare. In other words, it is not enough to provide care in the only objective knowledge of the disease, founded bio-medically on a mechanical theory of the body, which would stick to the chronicity of the disease and monitoring of a *temporalized subject* in accordance with protocol monitoring practices. Such a mechanical theory has been deemed unsatisfactory for more than 50 years, as evidenced by this comment from Canguilhem: "After being long accepted as a dogma by biologists, the mechanical theory of the body is today held by biologists claiming dialectical materialism as a narrow and inadequate view" [CAN 09, p. 129]. Responsible chrono-sensitive telecare aims to equally recognize, and take charge of, caring for the chronic disease in its temporality and the psycho-emotional well-being of the person (if that is their wish). Such caring underlies and enriches the doctor–patient relationship. Doctor Jean-Michel Crabbé also stresses that medicine should not neglect the effects of psychology on physiology and the rhythms of the sick person (see Chapter 7):

> *With their patients, the doctor may hesitate on the most appropriate attitude: performing difficult and expensive examinations, or start with an emotional and instinctive assessment, interviewing each patient on their lifestyle, their sleep and their dreams. Medicine should not overlook the impact of emotional factors, the rhythms of life and sleep on hormonal and neurovegetative functions, and the rhythms of the whole body. All of these findings open the door to a vast area. Physiology and rhythms are subject to the influence of a neuropsychic system. One might wonder if the imbalances of this system are not primum movens of multiple pathologies* [CRA 15].

Chrono-sensitive telecare thus asserts itself as the device of a medicine that is less *personalized* than *personalizing*, i.e. fundamentally preventive medicine centered on the person recognized and recognizing themself as ailing unique subjectivity; medicine that develops in the balance between the relational and temporal care values, mentioned previously; medicine whereby the patient can prove to be a person in respect of ethical values, brought jointly by actors and the chrono-technology objects put at their disposal; medicine of which care ethics, inspired by the work of Carol Gilligan; and, to a lesser extent, ethical trends of personalism, can illuminate the scope in the context of telecare. However, before reaching this analysis, several ethical principles deserve to be reviewed.

## 8.4. Ethical foundations of chrono-sensitive telecare

The temporalities of the disease and of the person, followed on a daily basis as part of doctor–patient information sharing (see Chapter 5); the relationship between the patient (as co-caregiver) and the doctor; the sensitivity of the person of whom care technologies are intended to support the psycho-emotional reality of (even if progress remains to be made, as we have seen): these are the areas covered by telecare in terms of patient perception as a *temporal* and *tested individual*. In addition, as attested in certain protocols, telecare does not exclude, *in principle* if not in practice, real singularity, history and the hopes and fears of the person. The recharacterization of the patient as co-caregiver, the recognition of affects and their impact on the disease and the care given to the person beyond the therapeutic gesture in its technical nature, of which *care ethics* allows us to recognize the virtue and the positive and creative functions (if one understands care as a practice of carrying co-created and shared knowledge), are tangible evidence of the chrono-sensitive *trend* in telecare today, asserting itself in telemedicine. As Frédéric Worms highlights concerning care (this comment also applies to telecare): "What is meant by 'care' [in *care ethics*], is not only *a relief* in terms of leaving this term to the more or less appropriate response to a purpose conceived as more or less restrictive. It is in fact this relief when it takes the ever-increasing form [...] of technical and clinical knowledge, on the one hand, and of a relationship that is not negative but positive, even creative, on the other hand" [WOR 10]. We must, however, cautiously refer to the "chrono-sensitive" trend of telemonitoring, because the already experimental nature of the existing chrono-sensitive telecare protocols call for some caution as the possibility of their democratization is far from assured. Some initiatives in this sense, listed in this book, are models to follow. To support this movement, it seems relevant, on the basis of the evidence underpinning the legitimacy of such initiatives, to explore the foundations of care ethics that may apply to the chrono-sensitive telecare of chronic patients today.

### 8.4.1. *Autonomy and humanity of telecare*

Chrono-sensitive telecare, centered on the person, does not raise, in the first place, ethics of *pastoral*[3] autonomy – the patient *taking charge* under the leadership of a medical authority – underlined by a rational and functional body conception. It is not an ethics that would seek to legitimize practices placing the body, as a *technical* object of care, at the center of a skill network, of which the patient would be the ultimate node. Rather, chrono-sensitive telecare is an ethics of the person who regards the body as a technique in the philosophical sense of the term; in other words, a body that makes use of the technique as an extension of skills, as the philosopher Isabelle Queval discusses here, in the context of a care relationship placing vulnerability at the center of medical attention.

> *The body is technical once its skills are taken into account. It is because of its environment or the ever greater proximity, or the fusion, with artificial materials that make up prostheses and microchips. It is, finally, when fantasy extrapolates technical reality to project a completely artificial production of the body and, at the heart of this project, an "intelligence" production, supposedly material* [QUE 08].

Far from the defined "technical body" within the engineering sense of the term (ethics of autonomy), the ethics of the person in the context of chrono-sensitive telecare questions the humanity of the technique, the "humanly technical" body; a kind of "exodarwinism" in the sense that is conferred by Michel Serres to this expression:

> *I call exo-darwinism this original movement of bodies to objects that outsource the means of adaptation* [SER 01].

In the specific context of chrono-sensitive telecare, the defibrillator, the cycler and the insulin pump do not dehumanize the person; they provide support to a failing biological function in restoring it (temporarily) – they support the person's (physical and mental) efforts to maintain good health – so that it is possible to share here the questioning of Michel Serres targeting the so-called principle of dehumanization of the technique (here in a telemedical context):

> *How would technology dehumanize humans when it extrapolates the best body feature* [SER 01]?

3 The meaning given to this term by Nikolas Rose. See, concerning this point, note 30, p. 127.

This humanistic design of the technique refers to the theory of technique as an extension and as a mediation, inherited from Ernst Kapp. Bernadette Bensaude-Vincent summarizes the crux of the matter:

> *Kapp attaches the tool to the body by a triple bond: the tool extends the body, it imitates the shape of a body, finally it increases the power of the individual. In his eyes, the progress of techniques translates into the better handling of tools, leading to continuity further marked between the body and the tool. So the technical evolution goes in the direction of a more and more advanced integration of artifacts and the body, a cyborg before the term existed. Contrary to the modern cyborg, an individual equipped with Kapp is not really a hybrid in ontological terms. But at the epistemological level, in the order of knowledge, humans can only understand themself and come to self-awareness through the mediation of artifacts* [BEN 09].

The relationship between people and technology in the context of remote care does not make the individual a hybrid, in the sense given to this term by Bruno Salgues: "a new type of person, a mixture of two bodies combined into one. One of them is natural, the other is a product of technology" [SAL 16]. The person is neither a robot with a human face nor an "engineered" human. Material to become shaped by transhumanist[4] and post-humanist[5] ideals, the humanity of the person is not to be found in their "digital double" either, that of their brain or their members designed by cutting-edge research[6]. In other words, the humanity of the technique, in the sense given to this expression by Isabelle Queval, is what the ethics of telecare, inspired by Serres and Kapp, takes into account along with the "technological" representation of the human, and the person at the center of care values:

> *Through externalization as virtuality, the technical body turns out to be a body connected inherently to techniques that match it, prolong it, repair it or expand its performances. Far from being alien to the self, this malleability reveals a singular relationship between nature and the artificial. The externalizing process specifies both the humanity of*

4 Transhumanism is an international cultural and intellectual movement advocating the use of science and technology to improve the human condition, including increasing physical and mental capacities in human beings. Source: https://en.wikipedia.org/wiki/Transhumanism.
5 Post-humanism is a thought current, emerging in the late 20th century, that deals with the relationship of human to technology (including biotechnology) and radical and inevitable change that this relationship has caused or may cause in the future. Source: https://en.wikipedia.org/wiki/Posthumanism.
6 See, on this point, the work of Stéphane Avril at the Saint-Étienne École des Mines in the field of biomechanics of soft tissue as well as the dual digital version developed on the German website "forshungzentrum julich", available at: www.fz-juelich.de.

> *the technique, otherness that proceeds from it, which must be questioned, and the intrinsic power of subjectivation revealed in mirroring* [QUE 08].

Humanity, but also otherness and the power of subjectivation: these are the three attributes of the process of externalization of the body operated by the technique at the center of the ethics of the person centered on autonomy in the context of chrono-sensitive telecare. It is important to highlight the conferred meaning of autonomy here: it is not a question of limiting dependencies but the rule of conduct a person gives to themselves, in the classical sense, since the Greek words *autos* and *nomos*, respectively, mean "self and law, the rule". This principle intends, particularly in the context of telemonitoring, to involve the patient both in the decision-making process and care protocol.

Otherness induced by the externalization process of the body by the technique results in the emergence of a double figure: on the one hand, the patient who reveals to themself – here is their power of subjectivation – and to others, including to the medical team, as co-person with a sense of responsibility thanks to telecare technologies that pair them; and, on the other hand, that of the physician as caregiver whose authority gives way to an increased empathy inspired by the recognition of the person's skills, mentioned earlier, in a context of technical care.

The ethics of the person centered on autonomy in the context of chrono-sensitive telecare is obviously not ethics of control in the Kantian or Rawlsian sense that focuses on the person as a moral agent able to choose, or change, a life plan, as philosopher Corinne Pelluchon points out:

> *The idea of the priority of right over good and the principle of an individual self constituted prior to the ends thay they choose, insist upon a volontarist conception of the person who is only bound to the ends and roles that they choose for themself. These roles may include other people, but the insistence of Rawls to define, following Kant, the self as a moral agent able to choose a life plan and change it, favors more a thought of control than a philosophy where the responsibility would limit freedom inside because the subject would not be considered independently from their specific duties. It contrasts with Levinas' philosophy where the responsibility for the other is first compared to freedom. Similarly, the value given to autonomy, understood as the ability of the subject to give themselves goals, has little incentive to recognize the given or non-chosen elements that characterize human life* [PEL 14].

Person-centered ethics recognizes that the person, as a subject, has the possibility of adjusting their responsibility of care to specific duties associated with the practice of care. The example of a person in independent daily home dialysis can bring a concrete example of this principle: it is certainly for the power to choose a life with flexibility afforded by the uses of the cycler, and to adopt to that end a proactive attitude toward care in a Rawlsian sense, but the technical mastery is not *an objective in itself.* The patient's responsibility is, indeed, invested in the gesture that must be made to achieve the life goal that they have established in the context of a partnership with the doctor, which is based on the production and the negotiated release of moral feelings (trust, respect, etc.) and care practices that constitute what may be termed a *moral economy*[7] of home telemonitoring. In other words, telecare is not a practice of autonomy and absolute control of the person in an exclusive relationship to telecare devices, or to a technique that the analysis of *affordances* (see Chapter 7) has shown to be, among other things, impossibly neutral. Telecare is a technical relationship that involves an agreed upon sharing of responsibilities with the medical team *on a daily basis*. It is a shared and harmonized remote monitoring of the disease, *by necessity*, implying an approach to the sick that cannot be purely technocratic. As Xavier Guchet suggests in *La médecine personnalisée* [Personalized medicine]: there can be no medicine exhausting care for the sole benefit of a *cure*[8]. However, ethics centered on the person and applied to chrono-sensitive telecare cannot limit its field of practice to issues of autonomy, humanity of the technique or the technical body. There are also singularly registered ethical principles at the heart of the telecare relationship, namely *solicitude*, *concrete interactions* or *proximity and distance*, in which care ethics, which is a way of thought seeking, among other things, to identify "a set of moral conduct that is based on pooling, caring of others [...] the experience of a relational self rather than the separation of subjective spheres and the cult of a distanced self, clearly separated from others" [BRU 11b], allows here to enlighten the analysis.

## 8.5. Care ethics tuned to telecare

Developed by American psychologist Carol Gilligan at the turn of the eighties, care ethics, as philosopher Fabienne Brugière stresses, questions "the borders of morality, the possibility of a new moral life experience which women, largely but not exclusively, find themselves to be otherwise trustees, or at least witnesses. A new voice emerges that requests to be problematized, brought to light. It inaugurates

7 In the sense of a "network of saturated values of affects that exist and function in relationship with the others", quoted in [DAS 95].

8 Donald W. Winnicott, in a work called "Cure" recently published in French (see [WIN 15]) defines a "*cure*" by emphasizing its difference with "care": "*Cure*, in the sense of treatment of the disease and its cause, tends to take precedence over the meaning of care (care, interest, attention)", in [GUC 16].

a problem that is simultaneously philosophical, psychological, sociological and political, that of *care*" [BRU 11a]. This new voice evoked by Brugière, although singular, echoed a larger and more radical movement, in recent history, from which multiple variants of care ethics have been able to see the day.

Released in the United States in 1982, *In a Different Voice* (1982) [GIL 82][9] marked a radical break with the medical ethics of the time and found itself at the origin of what is sometimes called the "*care revolution*", i.e. the "transformation of ethical reflection and practices induced in the rest of the world by this book and those that followed it" [NUR 10]. Fundamentally, the principle of *care*, as highlighted here by philosopher Frédéric Worms, raises a double meaning related to two separate realities of care, namely *feeling* and *activity*:

> *The notion of care refers [...] to two main registers of meaning: the first, which is the feeling, and can be expressed in different ways ("care for", "about"); the second, which is an activity, or even a job, is also diversified ("take care", "care giving", for example) [...] One would not understand "care ethics, however, unless you pointed out how they also seek to give to this feeling, on the one hand, a normative, decisive and central scope, and to this activity, on the other hand, a social as well as a critical and constitutive significance* [WOR 10].

It is important to add that *care* covers various meanings: attention to another (*care for*), care that is provided (*take care of*), the fondness that is shown (*care for*). Finally, *care* is inseparable from the notion of a relationship, interdependence, vulnerability and consideration that resonates in a singular way in the context of telecare since these notions should be entered through the prism of chrono-sensitive practices. Prioritizing the subject in *care ethics* aims at the same time to show that the limit of *curing*, in its technical nature, can be exceeded, as can be the classic opposition between *care* and *cure*; precisely by the fact that *care* is an inclusive approach, in the context of telecare, which aims to make it *possible* through *probable (cure)* techniques, to realize the self project, mentioned previously[10]. *Care* is the art of the "taking care of" in a context of vulnerability and takes account of prevention, pre- and post-traumatic phases and the valence principle (see definition, pp. 206–207). Therefore, it is essential to understand telecare as a *technicized relationship*, sensitive to the temporality of disease and the patient, as *interdependence by interposed communicating objects*, as *co-managed vulnerability* in a programmed temporality and as *mutual consideration*, favored by the use of

9 Published in France in 1986 in a translation by Annick Kwiatek, Gilligan's book was then republished in [GIL 08].
10 See section 7.10 of this book.

chrono-technologies. The ethics of telecare entered through the prism of *care ethics* is thus a *techno-ethics* approach adjoining itself to different contemporary ethics stemming from care ethics.

As philosopher Vanessa Nurock emphasizes, original *care ethics* have, in effect, greatly diversified in recent decades:

> *[Carol Gilligan's approach to care] found itself at the center of debates concerning the "second wave of feminism" characterized by a radical critique of patriarchy, giving rise to a large number of comments in various academic disciplines. It has thus enabled the birth of a trend of moral and political thinking integrating varied views, so that we speak today not only about Carol Gilligan's care ethics but the ethics and policies of care in the plural* [NUR 10].

The possibility of adapting *care ethics* to chrono-sensitive telecare arises here in the wake of the reflection proposed by Joan Tronto in *Moral Boundaries: a Political Argument for an Ethic of Care* [TRO 93]. As Frédéric Worms highlights, Joan Tronto, in *Moral Boundaries*, "insists above all on the social and political dimensions of concrete activity to which the sense of solicitude gives rise, and on the importance to recognize it. It is important in this respect [...] to withdraw *care* from the 'relationship' and also from 'face-to-face'. It's a whole social life that is based on concrete support of vital vulnerabilities of men" [WOR 10]. The choice has, indeed, been taken here to separate the study from the exclusive link, mentioned earlier, between the issues of *care* and *gender* (care ethics is "feminine because of feminism", and not the other way around, says Carol Gilligan), but also, contrary to Joan Tronto, to set aside the political dimension of *care*, and this for two main reasons: on the one hand, the (feminist) political question of care ethics, particularly due to the principle of justice, largely exceeds the conceptual framework of this book. On the other hand, the *depoliticization* of original *care ethics*, as well as its degenerification, pave the way for a reinterpretation of principles of caring, concrete interaction and closeness and distance, at the heart of care, necessary for grasping the ethical issues of contemporary chrono-sensitive telecare.

A comment is imperative at this stage of the study. Tronto indeed criticizes Gilligan's care ethics to limit care to an interindividual relationship (to a dyadic relationship of care) not taking into account health institutions. However, for Tronto, *care* relates to any activity or practice taking the form of treatment or protection; hence her expanded definition of care:

> *[...] a generic activity which includes everything we do to keep, maintain and repair our "world", so that we can live in the best way possible. This world understands our bodies, ourselves and our*

> *environment, all elements that we seek to connect into a complex network, in support of life* [TRO 93].

Telecare, as technicized care, is *care* in a general sense, not to say universalist, given here by Tronto. However, at this stage of the development of telemedicine in France, and given the current level of ethical subdetermination of telecare, it seems risky to offer a *social ethics of telecare* today. Maybe this work will be conducted in the coming years. For now, let us accept to stick with telecare centered on an interindividual relationship. It is initially grounded in the ethical principles of previously mentioned medical ethics (see Chapter 2)[11]. Beyond the boundaries of medical ethics, three ethical principles of *care* are also relevant to guide the practices of chrono-sensitive telecare today.

What is the ethical conception of solicitude as an exercise of attention to needy and weakened others in the context of chrono-sensitive telecare?

Solicitude refers to the importance of the sphere of sensitivity – which was discussed in Chapter 3[12], and the provisions to the person that it can build. The provision for care, in *care ethics*, is not conceived as a provision separate from the sphere of the activities of daily life nor as a specifically female ability. Similarly, the provision of care, or solicitude, refers to behaviors that involve the sphere of feelings and "emotional" intelligence. As Brugière also highlighted:

> *It requires questioning the philosophical development throughout the history of a* homo sentimentalis, *which means claiming a positioning on the stage of the history of philosophy on the side of Scottish (mainly Hume and Smith) and French (Diderot, Rousseau) enlightenment, the deployment of a certain skepticism about the abstract reason, of an experimental analysis of the natural power of feelings as a driver of activity. However, one must not assume any priority or purity from the provision. The latter is always built in social determinations and reorganized by activity that always transforms it in return* [BRU 11b].

*Homo sentimentalis*, in the context of telecare, is the caregiver. Their solicitude is an integral part of the relational dimension of telecare, whose ethical scope has been assessed here. The exercise of the caregiver's attention to the person is motivated by medical reasons as well as by a provision for care, more or less natural, that needs to be nurtured, from which can be put in place, among other

11 See Chapter 2.
12 See Chapter 3.

therapeutic actions, active listening and psychological support for the person[13]. The provision for care in the context of telecare, or the requalification of the concept of solicitude, which is central to care ethics, marks a break with orderly practice and prescriptive medicine (reviewed in Chapter 6). Also, the exercise of solicitude and the resulting provisions stand out from contemporary telemedicine practices that are deemed impersonal. It ultimately underlies, as demonstrated by patient testimonials, a relational practice of telecare, which still has to be improved, at the center of the current ethical debate on telemedicine.

QUESTION 8.1.– What are the concrete interactions in the context of telecare and how, from *care*, can one measure their ethical scope?

In the central chapter on "notions of self and morality" in *In a Different Voice*, Gilligan shows how analysis of "the place of women in the life of men" and the "images of the relationship" are implemented in practical decisions, in this case, that of abortion. Gilligan basically shows how "investigation into the decision to abort suggests that women build moral issues differently [from men] in terms of conflicting responsibility" [GIL 82]. What should be remembered here, beyond argument, is the fact that the practical decision is rooted in a concrete social reality and not an abstract concept of the individual and the care relationship. This is, therefore, the way in which the decision of care is taken as much as, it seems, as the concern for others that matters. Or as Worms emphasizes:

> *The decision will be made not on the basis of a general principle of an independent subject, but based on its effects on concrete relationships in which the subject who develops it is taken. This is the whole moral reasoning that is thus transformed, from the relational perspective to which the listening of the "different voice" conducted by Carol Gilligan is finally led* [WOR 10].

The way to make a care decision based on specific interactions, in the sense that Gilligan envisioned, is to focus on the relational perspective in the decision-making process, which is one of the ethical principles of care, returning to the concept of a co-caregiver of telecare: they cannot make a decision – injecting a dose of insulin in the case of a person with diabetes, for example – through communicating objects and self-care devices – without informing, or being informed by, the prescribing doctor. The decision, even taken remotely by electronic means, is therefore shared between the doctor and the patient in advance and this is because it is from this decision that the patient can perform home care safely by themselves. The ethical principle of concrete interaction is therefore respected here. It makes the relational

13 See Chapter 6.

dimension essential as a *sine qua non* condition to the decision of care and highlights the implicit ethical role of communicating objects mentioned above.

Following the example of care, although in different ways, the ethics of the chrono-sensitive telecare are also ethics of proximity and distance[14]: the various stages of the subjectivation of the patient-as-person, or ailing unique subjectivity, are relational variants of this dialectic whose essential features are highlighted by Gilligan here:

> *We know ourselves apart from others only to the extent that we live with them, and we feel connected to others only to the extent that we distinguish others from ourselves* [GIL 82].

In other words, the evolution of the "self" – as patient, person, then co-caregiver – is an outcome of the telecare relationship – which is constructed with regard to health status – whose criterion in the context of telecare is to empower the person by delegating them the temporal and communicational responsibility of care (through the transfer of their personal data, in particular), but also offering them the opportunity to share this experience with the health care team in real time. This vital link with the medical team is, however, not uniquely or mundanely a practical modality of telecare: it is also, from an ethical point of view, an objective technology of care with which subjective attention to the person can and must be maintained, sharing in this the aim of care analyzed here by Worms:

> *The subjective attention creeping into the objective or "technical" practices of care, or in the actions that treat a disease or prevent a danger as such, is the very condition of care, not only care of something but also care of someone. It is only when it is addressed that care constitutes subjects, but, conversely, it is also only if this address [...] is practiced effectively in an objective and concrete treatment that it will constitute these subjects, which are first of all physical individualities and living, real psyche, in the world* [WOR 10].

The symmetry of the effects of the dialectic of proximity and distance, in the doctor–patient relationship in medicine and telemedicine allows us to justify here

14 Proximity and distance, in *care* ethics, refers to the dialectic that integrates both initially opposed perspectives, feminine and masculine, the second being finally grasped and integrated theoretically and socially to the first. This "cross-pollination" of viewpoints refers, however, anaphorically, to the confrontation from the viewpoint of the caregiver with that of the care recipient in the decision-making framework of a telecare protocol.

the relevance of the convening of *care* ethics, in addition to the fundamentally ethical value of the relational dimension, between proximity and distance involved in telecare.

### 8.5.1. *Normative evaluation of the chrono-sensitive telecare*

Current chrono-sensitive telecare practices are rooted in the conception of a relational, interdependent and vulnerable individual whose temporal regime (based on their experience of the illness and the temporality of independent home care) as well as their psycho-emotional dimension can now be jointly supported, through chrono-health technology, as part of care ethics and self-concern, or *care ethics* revisited in the light of techno-medical progress.

In brief, there are multiple contributions and benefits of an ethical approach to *care* applied to chrono-sensitive telecare. This approach redefines, firstly, the sense of autonomy-related responsibility of the telecared person: in this context, the delegation of responsibility for patient care is less an end than a means. Assigning the person to live face-to-face with a machine, under the eye of a distant but vigilant authority – at the center of a "digital" panopticon – cannot be the ultimate goal of telecare. This device must be regarded as a mere *support* of care. Indeed, independent home care practices and patient testimonies show that the therapeutic alliance forged at the technical and biological crossroads calls, realistically, for a strengthening of the links between the caregiver and the care recipient and an increased humanization of protocols[15]. In addition, two key concepts of care ethics are applied here to telecare: solicitude and the taking into account of concrete interactions.

The ethics of chrono-sensitive telecare aims, ultimately, to oversee an approach of digital remote care that is holistic and benevolent; a form of care that is centered on the person as a co-caregiver exercising an independent ability through ethical and chrono-technological agents. This approach similarly tends to enhance greater care of the chrono-sensitive dimension in the telecare relationship and encourages an empathetic approach to the care provided to the person. Ultimately, telecare ethics stands out in reviving the special purposes of *care* ethics: to break the loneliness of the individual, to contribute to the reunification of their experience and their identity, to support the construction of the sense that they provide to the test of chronic diseases, and to support them in order to bring about genuine autonomy.

---

15 This situation created a kind of economic paradox: the cost reduction achieved through telemedicine (drop in hospital attendance, enhanced prevention, harmonization of monitoring) indeed seems counterbalanced today by the need to strengthen human resources in the psychological support of telecare (most notably in the case of chronic diseases).

# Conclusion

## Reconciling the Humanity and Technicality of care

*The value of life, life as a value, does it not take root in the knowledge of its essential insecurity?*

Georges Canguilhem[1]

Praised for its effectiveness or scorned for its technicality, telecare, which is one of the flagship mechanisms of telemedicine, have brought in, in less than a decade, conventional care and the doctor–patient relationship into a new era: one of digital health. Now inevitable, and actively supported by health authorities, telecare – i.e all the means of monitoring and practices of care that can be managed independently at home by chronic patients – upsets the therapeutic alliance that traditionally unites the patient to the doctor. The transformation is first of lexical order: we no longer talk about the *patient*, but the *telepatient* and even *the interactive patient* today. But who are the people concerned by telecare? It is, as shown in the analysis proposed in this book, the personal being as co-caregiver whose knowledge about their disease, their temporal regime and their psycho-affective situation must now be cared for by telecare.

The advent of telemedicine and telecare also changes the scale of values. The biological order of the patient at the center of the *ordered* model of medicine can no longer be the unique mark of medical practices. The biological and the psychological nature of the patient must be jointly supported – especially in the context of care under self-managed home telemonitoring care – in order to address

1 Source: [CAN 02], http://www.universalis.fr/encyclopedie/vie/. Page consulted on March 13, 2015.

the human ethical challenge that is beyond the scope of a strictly technical doctor–patient relationship, mediated through interposed objects.

Viewed from a technical perspective, telecare seems to be an objectifying practice in which the patient is perceived as a neutral support of a symptom to be deciphered and as a figure that evolves over time and differently according to the observed pathology: data producer in real time (cardiac patient), chrono-empowered patient (renal failure) or synchronized interactive patient (diabetic patient). The patient experience is limited to a cold and distant relationship with the doctor, and to daily interactions, devoid of all humanity, with technical care objects.

Viewed from the perspective of people with chronic diseases, the practice of telecare is however at odds with the label of impersonal medicine that we attach to it sometimes. Of course, as has been pointed out repeatedly in this book, the attention to the patient, their psychological support and humanity of the care in general, are far from being acquired at the current stage of development of telecare in France. Additional human resources should be mobilized, in addition to financial resources, to complete the medical teams actively involved in the success of telecare. This passes notably, as has been said, by taking into account the person, in the sense given by the present research to this term, and a recognition of their knowledge, their experience and their feelings of illness and self-managed home care. This recognition is already acquired in a number of protocols and the work of medical teams is already going in the sense of a strengthening of the link with the patient and a tight mesh of various actors in care, including outside the protocol (city doctor pharmacist, patient associations). The existing good practices that allow us to improve significantly, or even considerably in some cases, the management of the person, in all their diversity, in the context of telecare and therefore their quality of life, are however local or regional initiatives, which are in demand of an ethical assessment.

The reflection in this book leads to clear ethical principles to guide and regulate the practices of telecare for people today with kidney failure, heart failure and diabetes. Based on the testimonies of these individuals, the principles of humanized care and autonomy – foundations of an ethical edifice that remains to be consolidated through large scale qualitative studies integrating the temporal and psycho-emotional parameters of the patients – compel us to redefine, in the first place, the value that should be given to the *patient-as-person* in the remote caregiver–care recipient relationship. These principles also imply we should recognize the moral agency and the chrono-technological function of telecare objects – constitutive of the chrono-sensibility concept according to which temporality and relatedness cannot be separated in a telecare context. Finally, these principles point to the necessity of jointly taking into account the medical, temporal and relational dimensions of self-managed care.

Thus defined, telecare ethics ultimately show that a chronic disease most notably cannot be understood as a malfunction but as a precarious state which is also, according to Canguilhem, that of life and that which gives to life, therefore to the disease, its meaning and its value. Vulnerability is a way of being that involves a redeployment of the subject in which telecare, beyond its technical nature, must take part. Digital clocks that govern the care are robots put to the service of a treatment whose ethical value can indeed be preserved only as long as you protect the property the most precious to the person, patient or doctor: their humanity.

# List of Abbreviations

## A

**ADP**: Automated peritoneal dialysis

**AGGIR**: Autonomie gérontologique groupe iso-ressources (a metric to assess the level of independence among the elderly)

**ANAP**: *Agence nationale d'appui à la performance hospitalière* (French national agency for hospital performance Support)

**ANSM**: *Agence nationale de sécurité du médicament et des produits de santé* (French national agency for the safety of medicines and health products)

**ANTEL**: *Association nationale pour la télémédecine* (French national association for telemedicine)

**AP-HP**: *Assistance publique – hôpitaux de Paris* (Public assistance – Paris hospitals)

**APODEC**: *Association de porteurs de défibrillateurs cardiaques* (French cardiac defibrillator carrier association)

**ARC**: Academic research community

**ARS**: *Agence régionale de santé* (French regional health agency)

**ASIP santé**: *Agence des systèmes d'information partagés de santé* (French shared health information systems agency)

## C

**CALYDIAL**: *Centre associatif lyonnais de dialyse* (Dialysis association center of Lyon)

**CATEL**: *Club des acteurs de la télésanté* (French telehealth stakeholders club)

**CBGM**: Continuous blood glucose monitoring

**CCPD**: Continuous cycling peritoneal dialysis, also known as automated peritoneal dialysis (APD)

**CERITD**: *Centre d'études et de recherche sur l'intensification du traitement du diabète* (French center for diabetes treatment intensification studies and research)

**CES**: Computer engineering services company

**CHF**: Chronic heart failure

**CHSF**: *Centre hospitalier Sud Francilien* (South Francilien Hospital Center)

**CKD**: Chronic kidney disease (also known as chronic renal failure or chronic renal insufficiency)

**CNIL**: *Commission nationale de l'informatique et des libertés* (French national commission on informatics and liberties)

**CNOM**: *Conseil national de l'ordre des médecins* (French national medical association council)

**CNRS**: *Centre national de la recherche scientifique* (French national center for scientific research)

**CRUQ**: *Commission des relations avec les usagers et la qualité de la prise en charge* (French commission on relations with users and the quality of management)

**CVA**: Cerebrovascular accident

## D

**DATAR**: *Délégation interministérielle à l'aménagement du territoire et à l'attractivité régionale* (French interministerial delegation for regional land-use planning and regional attractiveness)

**DGCIS**: *Direction générale de la compétitivité de l'industrie et des services* (General management of industry competitiveness and services)

**DGCS**: *Direction générale de la cohésion sociale* (General management of social cohesion)

**DGOS**: *Direction générale de l'offre de soins* (General management of healthcare provision)

**DNA**: Deoxyribonucleic acid

**DSSIS**: *Délégation à la stratégie des systèmes d'information* (Information systems strategy delegation)

## E

**EBM**: Evidence-based medicine

**EFT**: Episodic future thinking

**EHPAD**: *Établissements d'hébergement pour personnes âgées dépendantes* (Residential facilities for dependent elderly persons)

**EPO**: Erythropoietin

**ESPIC**: *Établissement de santé privé d'intérêt collectif* (Private health establishment of collective interest)

**ESRD**: End-stage renal disease

## F

**FBG**: Fasting blood glucose

**FDA**: Food and Drug Administration

**FEHAP**: *Fédération des établissements hospitaliers et d'aide à la personne* (Federation of hospitals and personal care institutions)

## G

**GCS**: *Groupement de coordination sanitaire* (Health coordination group)

**GSM**: Global system for mobile communication

## H

**HAD**: Hôpital à domicile (Hospital at home)

**HAS**: *Haute autorité de santé* (French high authority for health)

**HBP**: High blood pressure

**HPST**: *Loi hôpital, patients, santé, territoires* (Hospital, patients, health, territories law)

## I

**ICD**: Implantable cardioverter defibrillator

**ICT**: Information and communication technologies

**IHH**: In-home hospitalization

**IMD**: Implantable medical devices

**INPES**: *Institut national de prévention et d'éducation pour la santé* (National institute for prevention and health education)

**INRIA**: *Institut national de recherche en informatique et en automatique* (French national institute for research in computer science and automation)

**INSEE**: *Institut national de la statistique et des études économiques* (French national institute of statistics and economic studies)

**INSERM**: *Institut national de la santé et de la recherche médicale* (French national institute of health and medical research)

**INVS**: *Institut de veille sanitaire* (Health monitoring institute)

**IS**: Information system

## L

**LDL**: Low-density lipoproteins

**LFSS**: *Loi de finances de la sécurité sociale* (French social security finance law)

**LTC**: Long-term condition

## M

**MAST**: Model of assessment for telemedicine

**MDM**: Mobile device management

**MRI**: Magnetic resonance imaging

## N

**ND-CKD**: Non-dialysis chronic kidney disease

**NSP**: National strategic plan

## O

**OCDE**: *Organisation de coopération et de développement économiques* (Organization for economic cooperation and development)

**OPIIEC**: *Observatoire paritaire des métiers de l'informatique, de l'ingénierie, des études et du conseil* (Joint observatory of IT, engineering, research and consulting professions)

## P

**Pa**: Pascal (symbol for the unit)

**PACA**: Provence-Alpes-Côte d'Azur region of France

**PGI**: Patient-generated index

**PHC**: Professional health card

**PMR**: Personal medical record

**PPG**: Postprandial glycemia

**PRS**: *Projet régional de santé* (French regional health project)

**PRT**: *Programme régional de télémédecine* (French regional telemedicine program)

**PSH**: Personalized self-help

**PSPH**: *Établissements participant au service public hospitalier* (Establishments participating in the French public hospital service)

## R

**R&D**: Research and development

**RFID**: Radio frequency identification

**RMHMS**: Remote medical home monitoring system

**RN**: Registered nurse

**RTP**: Regional telemedicine program

## S

**SCAD**: *Suivi clinique à domicile* (Clinical follow-up at home for cardiac patients)

**SCN**: Suprachiasmic nuclei

**SEIQoL**: Schedule for the evaluation of individual quality of life

**SMS**: Short message service

**SNITEM**: *Syndicat national de l'industrie des technologies médicales* (French national union of the medical technologies industry)

**SQLP**: Subjective quality of life profile

**SYNTEC**: *Fédération française des syndicats professionnels spécialisés dans les professions de l'ingénierie, du numérique, des études et du conseil, de la formation professionnelle et de l'événementiel* (French federation of professional unions specializing in the engineering, digital, research and consulting, professional and event training)

## T

**T1D**: Type 1 diabetes

**T2D**: Type 2 diabetes

**TEMSED**: Evaluation method "technology, ergonomics, medical, social, economic, deontology"

**TH**: Thyroid hormone

**TSH**: Thyroid-stimulating hormone

**U**

**UHC**: University hospital center

**UNCANSS**: *Union des caisses nationales de sécurité sociale* (French union of national social security funds)

**UNESCO**: United Nations Educational, Scientific and Cultural Organization

**UNO**: United Nations Organization

**URL**: Uniform resource locator

**URPS**: *Union régionale des professionnels de santé* (French regional union of health professionals)

**W**

**WHO**: World Health Organization

# References

## Books


[ADA 98] ADAM B., *Timescapes of Modernity: The Environment and Invisible Hazards*, Routledge, London, 1998.

[AGA 09] AGAMBEN G., *Nudités*, Éditions Payot & Rivages, Paris, 2009.

[AND 01] ANDERS G., *L'obsolescence de l'homme*, Ivréa, Paris, [1956], 2001.

[AND 11a] ANDERSON M., ANDERSON S.L. (eds), *Machine Ethics*, Cambridge University Press, Cambridge, 2011.

[AND 11b] ANDRIEU B., *Les avatars du corps. Une hybridation somatechnique*, Liber, Montreal, 2011.

[ARE 58] ARENDT H., *La Condition de l'homme moderne*, Folio essais, Paris, 1958.

[ARE 72] ARENDT H., *La Crise de la culture*, Folio essais, Paris, 1972.

[ARI 08] ARISTOTLE, *Sur la justice. Éthique à Nicomaque, Livre V*, GF Flammarion, Paris, 2008.

[BAC 92] BACHELARD G., *L'intuition de l'instant*, Éditions Stock, Paris, [1931], 1992.

[BAC 13] BACHELARD G., *La dialectique de la durée*, PUF, Paris, [1939], 2013.

[BAC 14] BACRO F. (ed.), *La qualité de vie, Approches psychologiques*, Presses universitaires de Rennes, Rennes, 2014.

[BAR 13] BARRUEL F., BIOY A. (eds), *Du soin de la personne, clinique de l'incertitude*, Dunod, Paris, 2013.

[BAR 14a] BARRIER P., *Le patient autonome*, PUF, Paris, 2014.

[BEA 08] BEAUCHAMP T., CHILDRESS J., *Les principes de l'éthique biomédicale*, Les Belles Lettres, Paris, 2008.

[BEN 03] BENSAUDE-VINCENT B., STENGERS I., *Cent mots pour commencer à penser les sciences*, Les Empêcheurs de Penser en Rond, Paris, 2003.

[BEN 09] BENSAUDE-VINCENT B., *Les vertiges de la technoscience. Façonner le monde atome par atome*, La Découverte, Paris, 2009.

[BEN 11] BENSAUDE-VINCENT B., BENOÎT-BROWAEYS D., *Fabriquer la vie: Où va la biologie de synthèse ?*, Le Seuil, Paris, 2011.

[BEN 13a] BENSAUDE-VINCENT B., *L'opinion publique et la science. À chacun son ignorance*, La Découverte, Paris, 2013.

[BER 05] BERGSON H., *La pensée et le mouvant*, PUF, Paris, [1938], 2005.

[BER 09] BERNS T., *Gouverner sans gouverner. Une archéologie politique de la statistique*, PUF, Paris, 2009.

[BER 12] BERGSON H., *Matière et mémoire*, Flammarion, Paris, [1896], 2012.

[BER 13] BERGSON H., *Les données immédiates de la conscience*, Flammarion, Paris, [1888], 2013.

[BER 15a] BERTHOZ A., DEBRU C. (eds), *Anticipation et prédiction. Du geste au voyage mental*, Odile Jacob, Paris, 2015.

[BÉR 16] BÉRANGER J., *Big Data and Ethics: The Medical Datasphere*, ISTE Press Ltd, London and Elsevier, Oxford, 2016.

[BIR 12] BIRTH K.K., *Objects of Time. How things Shape Temporality*, Palgrave Macmillan, New York, 2012.

[BRI 08] BRISSON L., *Platon. Œuvres complètes*, Flammarion, Paris, 2008.

[BRU 11a] BRUGIÈRE F., *L'éthique du care*, PUF, Paris, 2011.

[BUS 11] BUSER P., DEBRU C., *Le temps, instant et durée. De la philosophie aux neurosciences*, Odile Jacob, Paris, 2011.

[CAN 02] CANGUILHEM G., *Écrits sur la médecine*, Le Seuil, Paris, 2002.

[CAN 10] CANGUILHEM G., *Le normal et le pathologique*, PUF, Paris, [1966], 2010.

[CAS 05] CASSIRER E., *Le problème de la connaissance dans la philosophie et la science des Temps modernes*, vol. 2, Le Cerf, Paris, 2005.

[CHA 07] CHANGEUX J.-P. (ed.), *L'Homme artificiel*, Odile Jacob, Paris, 2007.

[DAM 94] DAMASIO A., *Descartes' Error: Emotion, Reason, and the Human Brain*, Grosset/Putman, New York, 1994.

[DEL 90] DELEUZE G., *Pourparlers*, Éditions de Minuit, Paris, 1990.

[DES 08] DESROSIÈRES A., *Pour une sociologie historique de la quantification. L'argument statistique I*, Presses de l'École des mines, Paris, 2008.

[DOU 11] DOUEIHI M., *Pour un humanisme numérique*, Le Seuil, Paris, 2011.

[DUB 01] DUBEY G., *Le lien social à l'ère du virtuel*, PUF, Paris, 2001.

[DUM 04] DUMIT J., *Picturing Personhood: Brain scans and biomedical identity*, Princeton University Press, Princeton, NJ, 2004.

[FEE 04] FEENBERG A., *(Re)penser la technique: vers une technologie démocratique*, La Découverte, Paris, 2004.

[FOU 01a] FOUCAULT M., *Dits et Écrits II.1976-1988*, Gallimard, Paris, 2001.

[FOU 01b] FOUCAULT M., *L'herméneutique du sujet. Cours au Collège de France. 1981-1982*, Gallimard/Le Seuil, Paris, 2001.

[FOU 04] FOUCAULT M., *Naissance de la biopolitique*, Gallimard/Le Seuil, Paris, 2004.

[FOU 09] FOUCAULT M., *Naissance de la clinique*, PUF, Paris, [1963], 2009.

[GAD 93] GADAMER H.-G., *Philosophie de la santé*, Grasset-Mollat, Paris, 1993.

[GIB 79] GIBSON J.J., *The Ecological Approach to Visual Perception*, Houghton-Mifflin, Boston, 1979.

[GIL 82] GILLIGAN C., *In a Different Voice: Psychological Theory and Women's Development*, Harvard University Press, Cambridge, 1982.

[GIL 08] GILLIGAN C., *Une voix différente: pour une éthique du* care, Flammarion, Paris, 2008.

[GOF 08] GOFFETTE J., *Naissance de l'anthropotechnie. De la médecine au modelage de l'humain*, Vrin, Paris, 2008.

[GOL 01] GOLSE B., *Du corps à la pensée*, PUF, Paris, 2001.

[GON 01] GONORD A., *Le temps*, GF Flammarion, Paris, 2001.

[GUC 10] GUCHET X., *Pour un humanisme technologique. Culture, technique et société dans la philosophie de Gilbert Simondon*, PUF, Paris, 2010.

[GUC 16] GUCHET X., *La médecine personnalisée. Un essai philosophique*, Les Belles Lettres, Paris, 2016.

[HAB 90] HABERMAS J., *La technique et la science comme "idéologie"*, Gallimard, Paris, 1990.

[HAB 02] HABERMAS J., *L'avenir de la nature humaine*, Gallimard, Paris, 2002.

[HAL 01] HALPERN J., *From Detached Concern to Empathy. Humanizing Medical Practice*, Oxford University Press, Oxford, 2001.

[HAR 03] HARTOG F., *Régimes d'historicité. Présentisme et expérience du temps*, Le Seuil, Paris, 2003.

[HEI 58] HEIDEGGER M., *Essais et conférences*, translation by André Préau, Gallimard, Paris, 1958.

[HEI 86] HEIDEGGER M., *Être et Temps*, Gallimard, Paris, [1927], 1986.

[HIR 10] HIRSCH E. (ed.), *Traité de bioéthique*, vol. 1, Erès, Toulouse, 2010.

[ILL 88] ILLICH I., *Medical Nemesis. The Expropriation of Health*, Random House, New York, [1975], 1988.

[JON 98] JONAS H., *Pour une éthique du futur*, Éditions Payot & Rivages, Paris, 1998.

[KAN 97] KANT E., *Leçons d'éthique*, Librairie Générale Française, Paris, 1997.

[KAN 07] KANT E., *Écrits sur le corps et l'esprit*, GF Flammarion, Paris, 2007.

[KAP 07] KAPP E., *Principes d'une philosophie de la technique*, Vrin, Paris, [1877], 2007.

[LAF 04] LAFONTAINE C., *L'empire cybernétique*, Le Seuil, Paris, 2004.

[LAT 01] LATOUR B., *L'espoir de Pandore. Pour une version réaliste de l'activité scientifique*, La Découverte, Paris, 2001.

[LEB 92] LE BRETON D., *La sociologie du corps*, PUF, Paris, 1992.

[LEB 05] LE BRETON D., *Anthropologie du corps et de la modernité*, PUF, Paris, 2005.

[LEG 08] LEGUIL C., RUSS J., *La pensée éthique contemporaine*, PUF, Paris, 2008.

[LER 65] LEROI-GOURHAN A., *Le geste et la parole: La mémoire et les rythmes*, vol. 2, Albin Michel, Paris, 1965.

[LEV 95] LEVINAS E., *Alterité et transcendance*, Fata Morgana, Paris, 1995.

[LEW 00] LEWIS M., HAVILAND-JONES J.M., FELDMAN B.L. (eds), *Handbook of Emotions*, The Guilford Press, New York, 2000.

[MAC 09] MACHEREY P., *De Canguilhem à Foucault. La force des normes*, La Fabrique éditions, Paris, 2009.

[MAH 15] MAHÉO G., HOUSSET E. (eds), *Max Scheler: éthique et phénoménologie*, Presses universitaires de Rennes, Rennes, 2015.

[MAR 68] MARCUSE H., *L'Homme unidimensionnel*, Les Éditions de minuit, Paris, 1968.

[MCG 04] MCGRATH J.E., TSCHAN F., *Temporal Matters in Social Psychology: Examining the Role of Time in the Lives of Groups and Individuals*, American Psychological Association Publications, Washington DC, 2004.

[MER 64] MERLEAU-PONTY M., *Le visible et l'invisible*, Gallimard, Paris, 1964.

[MER 15] MERLEAU-PONTY M., *Phénoménologie de la perception*, Gallimard, Paris, [1945], 2015.

[MOU 49] MOUNIER E., *Le personnalisme*, PUF, Paris, 1949.

[NIN 13] NINOT G., *Démontrer l'efficacité des interventions non médicamenteuses: Question de point de vue*, Presses universitaires de la méditerranée, Montpellier, 2013.

[NOR 02] NORMAN D., *The Design of Everyday Things*, Basic Books, New York, 2002.

[OUD 11] OUDSHOORN N., *Telecare Technologies and the transformation of Healthcare*, Palgrave Macmillan, Basingstoke, 2011.

[PEL 14] PELLUCHON C., *L'autonomie brisée. Bioéthique et philosophie*, PUF, Paris, 2014.

[PEL 16] PELLÉ S., REBER B., *From Ethical Review to Responsible Research and Innovation*, ISTE Ltd, London and John Wiley & Sons, New York, 2016.

[PLA 17] PLATO, Complete Works, Hackett Publishing Company, Indianapolis, 1997.

[POL 12] POLS J., *Care at a Distance. On the Closeness of Technology*, Amsterdam University Press, Amsterdam, The Netherlands, 2012.

[QUE 08] QUEVAL I., *Le corps aujourd'hui*, Gallimard, Paris, 2008.

[REB 11] REBER B., *La démocratie génétiquement modifiée. Sociologies éthiques de l'évaluation des technologies controversées*, Presses de l'université Laval, Laval, 2011.

[RIC 85] RICOEUR P., *Temps et récit*, Le Seuil, Paris, 1985.

[ROS 07] ROSE N., *The Politics of Life Itself. Biomedicine, Power, and Subjectivity in the Twenty-First Century*, Princeton University Press, Princeton, NJ, 2007.

[ROS 10a] ROSA H., *Accélération. Une critique sociale du temps*, La Découverte, Paris, 2010.

[ROS 10b] ROSA H., *Aliénation et accélération. Vers une théorie critique de la modernité tardive*, La Découverte, Paris, 2010.

[ROU 08] ROUVROY A., *Human Genes and Neo-Liberal Governance. A Foucauldian Critique*, Routledge, New York, 2008.

[RUS 94] RUSS J., *La pensée éthique contemporaine*, PUF, Paris, 1994.

[SAL 11] SALANSKIS J.-M., *Le monde du computationnel*, Encre Marine, Paris, 2011.

[SAL 16] SALGUES B., *Health Industrialization*, vol. 1, ISTE Press Ltd, London and Elsevier Ltd, Oxford, 2016.

[SAL 13a] SALGUES B., *L'e-santé et la télémédecine*, Hermès–Lavoisier, Paris, 2013.

[SAL 13b] SALGUES B., PAQUEL N., *Télésanté, espoir du monde rural? Des objectifs à la construction des usages*, Michel Houdiard Editeur, Paris, 2013.

[SAR 72] SARTRE J.-P., *L'existentialisme est un humanisme*, Éditions Nagel, Paris, 1972.

[SAR 01] SARTRE J.-P., *L'être et le néant*, Gallimard, Paris, [1943], 2001.

[SAV 94] SAVATER F., *Éthique à l'usage de mon fils*, Le Seuil, Paris, 1994.

[SAV 95] SAVATER F., *Politique à l'usage de mon fils*, Le Seuil, Paris, 1995.

[SAV 00] SAVATER F., *Penser sa vie. Une introduction à la philosophie*, Le Seuil, Paris, [1999] 2000.

[SAV 05] SAVATER F., *Sur l'art de vivre*, Hachette Littératures, Paris, [1994], 2005.

[SEN 87] SEN A., *Commodities and Capabilities*, Oxford India Paperback, Oxford, 1987.

[SER 85] SERRES M., *Les cinq sens*, Grasset, Paris, 1985.

[SER 01] SERRES M., *Hominescence*, Le Pommier, Paris, 2001.

[SFE 95] SFEZ L., *La santé parfaite. Critique d'une nouvelle utopie*, Le Seuil, Paris, 1995.

[SHE 10] SHERRINGTON C.S., *The Integrative Action of the Nervous System*, Nabu Press, Charleston, [1906], 2010.

[SIC 02] SICARD D., *La Médecine sans le corps. Une nouvelle réflexion éthique*, Plon, Paris, 2002.

[SIÉ 14] SIÉROFF É., DROZDA-SENKOWSKA E., ERGIS A.-M. *et al.*, *Psychologie de l'anticipation*, Armand Colin, Paris, 2014.

[SIM 89] SIMONDON G., *L'individuation psychique et collective*, Aubier, Paris, 1989.

[SIM 05] SIMONDON G., *L'individuation à la lumière des notions de forme et d'information*, Millon, Paris, 2005.

[SIM 09] SIMON D., TRAYNARD P.Y., BOURDILLON F. *et al.*, *Éducation thérapeutique. Prévention et maladies chroniques*, Elsevier Masson, Issy-Les-Moulineaux, 2009.

[SIM 12] SIMONDON G., *Du mode d'existence des objets techniques*, Aubier, Paris, [1958], 2012.

[SIM 15] SIMONDON G., *Sur la psychologie (1956-1967)*, PUF, Paris, 2015.

[SUT 91] SUTTER J., BERTA M., *L'anticipation et ses applications cliniques*, PUF, Paris, 1991.

[TOU 15] TOURETTE-TURGIS C., *L'éducation thérapeutique, la maladie comme occasion d'apprentissage*, De Boeck, Paris, 2015.

[TRO 93] TRONTO J.C., *Moral Boundaries: A Political Argument for an Ethic of Care*, Routledge, New York, 1993.

[TUR 99] TUREK F.W., ZEE P.C., *Regulation of Sleep and Circadian Rhythms*, Éditions Marcel Dekker, New York, 1999.

[VAN 14] VAN RIALER J., *La nouvelle gestion de soi. Ce qu'il faut faire pour vivre mieux*, Éditions Mardaga, Brussels, 2014.

[VIA 13] VIAL S., *L'être et l'écran. Comment le numérique change la perception*, PUF, Paris, 2013.

[WAJ 15] WAJCMAN J., *Pressed for Time. The Acceleration of Life in Digital Capitalism*, University of Chicago Press, Chicago, 2015.

[WOR 15] WORMS F., BENAROYO L., LEFÈVE C. *et al.* (eds), *La philosophie du soin. Éthique, médecine et société*, PUF, Paris, 2015.

## Chapters in books

[ARE 07] ARENDT H., "The conquest of space and the stature of man", *The New Atlantis*, available at: https://www.thenewatlantis.com/docLib/TNA18-Arendt.pdf, pp. 43–55, 2007.

[BAC 14] BACRO F., FLORIN A., "Entre complexité et richesse: la diversité des défis liés à l'intérêt des chercheurs et des professionnels pour la qualité de vie", in BACRO F. (ed.), *La qualité de vie, Approches psychologiques*, Presses universitaires de Rennes, Rennes, France, pp. 7–14, 2014.

[BAR 09] BARRIER P., "Le temps du patient (chronique)", in SIMON D., TRAYNARD P.Y., BOURDILLON F. *et al.* (eds), *Éducation thérapeutique. Prévention et maladies chroniques*, Elsevier Masson, Issy-Les-Moulineaux, pp. 39–52, 2009.

[BAR 10a] BARRIER P., "Le soin comme accompagnement et facilitation de l'individuation avec la maladie chronique", in BENAROYO L., LEFÈVE C., MINO J.-C. *et al.* (eds), *La philosophie du soin. Éthique, médecine et société*, PUF, Paris, pp. 155–171, 2010.

[BIE 11] BIEWENER C., BACQUÉ M.-H., "Empowerment, développement et féminisme : entre projet de transformation sociale et néolibéralisme", in BACQUÉ M.-H., SINTOMER Y. (eds), *La démocratie participative. Histoire et généalogie*, La Découverte, Paris, pp. 82–83, 2011.

[BRU 11b] BRUGIÈRE F., "L'éthique du care: entre sollicitude et soin, dispositions et pratiques", in BENAROYO L. *et al.* (eds), *La philosophie du soin. Éthique, médecine et société*, PUF, Paris, pp. 69–86, 2011.

[CAN 09] CANGUILHEM G., "Machine et organisme", *La connaissance de la vie*, Vrin, Paris, pp. 129–165, [1965], 2009.

[COE 07] COENEN-HUNTER J., "Classifications, typologies et rapport aux valeurs", in BRIDEL P. (ed.), *Revue européenne des sciences sociales*, vol. XLV-138, pp. 27–40, 2007.

[FLO 11] FLORIDI L., "On the morality of artificial agents", in ANDERSON M., ANDERSON S.L. (eds), *Machine Ethics*, Cambridge University Press, Cambridge, pp. 184–212, 2011.

[GAR 13] GARGIULO M., ROSENBLUM O., "L'éloge de l'incertitude en consultation génétique", in BARRUEL F., BIOY A. (eds), *Du soin de la personne, clinique de l'incertitude*, Dunod, Paris, pp. 171–177, 2013.

[GAR 14] GARGIULO M., "Anticipation, prédiction et incertitude en médecine prédictive", in SIÉROFF É., DROZDA-SENKOWSKA E., ERGIS A.-M., MOUTIER S. (eds), *Psychologie de l'anticipation*, Armand Colin, Paris, pp. 173–192, 2014.

[MAR 70] MARCUSE H., "Industrialisation et capitalisme chez Max Weber", *Culture et Société*, Les Éditions de minuit, Paris, pp. 271–293, 1970.

[NUR 10] NUROCK V., "Avant-propos. Et si les poules avaient des dents?", in NUROCK V. (ed.), *Carol Gilligan et l'éthique du* care, PUF, Paris, pp. 9–18, 2010.

[PER 15] PERRIN D., "Asymétries dans le temps subjectif", in BERTHOZ A., DEBRU C. (eds), *Anticipation et prédiction. Du geste au voyage mental*, Odile Jacob, Paris, pp. 193–204, 2015.

[PIE 10] PIERRON J.-P., "Corps viellissant, corps vil?" in ANCET P. (ed.), *Le corps vécu chez la personne âgée et la personne handicapée*, Dunod, Paris, pp. 101–120, 2010.

[TAN 13] TANNIER C., "Éthique de l'autonomie et relation de soins", *Revue hospitalière de France*, vol. 550, pp. 63–66, 2013.

[WIN 15] WINNICOTT D.W., "Cure", in MARIN C., WORMS F. (eds), *À quel soin se fier? Conversations avec Winnicott*, PUF, Paris, pp. 8–25, 2015.

[WOR 10] WORMS F., "Le *care* selon Carol Gilligan et le problème du soin: relations vitales et morales", in NUROCK V. (ed.), *Carol Gilligan et l'éthique du* care, PUF, Paris, pp. 137–150, 2010.

## Journal Articles

[ABD 11] ABDULRAZAK B., GIROUX S., BOUCHARD B. *et al.*, "Towards useful services for elderly and people with disabilities", *9th International Conference on Smart Homes and Health Telematics (ICOST)*, Springer, Berlin, 2011.

[ADA 13] ADAIR R., WHOLEY D.R., CHRISTIANSON J. *et al.*, "Improving chronic disease care by adding laypersons to the primary care team: a parallel randomized trial", *Annals of Internal Medicine*, vol. 159, no. 3, pp. 176–184, 2013.

[ALB 12] ALBRECHT U., "Timing to perfection: the biology of central and peripheral circadian clocks", *Neuron*, vol. 74, pp. 246–260, 2012.

[ALS 13] AL-SHORBAJI N., "Is there and do we need evidence on eHealth interventions?", *IRBM*, vol. 34, no. 1, pp. 24–27, 2013.

[ANG 97] ANGRILLI A., CHERUBINI P., PAVESE A. *et al.*, "The influence of affective factors on time perception", *Perception and Psychophysics*, vol. 59, pp. 972–982, 1997.

[BAR 10b] BARTZ C., "International council of nurse and person-centered care", *International Journal of Integrated Care*, vol. 10, no. 5, pp. 24–26, 2010.

[BAR 15] BARDY P., "Patient empowerment: a practical approach to guide ethical decision-making in telemedicine", *Eä: Journal of Medical Humanities and Social Studies of Science*, vol. 7, nos 1–2, pp. 43–61, 2015.

[BEN 13b] BENHAMOU P.-Y., MULLER M., LABLANCHE S. *et al.*, "La télémédecine au service de la prise en charge des patients diabétiques: développements actuels et conditions du succès", *European Research in Telemedicine*, vol. 2, pp. 23–28, 2013.

[BEN 13c] BENSAUDE-VINCENT B., BARDY P., "Public engagement in science", *Alliages*, vol. 72, pp. 93–101, 2013.

[BER 15b] BERNARD M.-M., GALLON O., EL GHARBI T. *et al.*, "Evaluation by 30 chronically ill patients of a connection to their specialist physician, using a telemedicine-telementoring interface", *European Research in Telemedicine*, vol. 4, pp. 103–107, 2015.

[BRA 14] BRAMSTEDT K.A., PRANG M., DAVE S. *et al.*, "Telemedicine as an ethics teaching tool for medical students within the nephrology curriculum", *Progress in Transplantation*, vol. 24, no. 3, pp. 294–297, 2014.

[BUI 03] BUIN Y., "Normopathie", *Le Passant Ordinaire, Revue Internationale de Création et de Pensée Critique*, nos 45–46, pp. 44–46, 2003.

[CAI 14] CAILLETTE-BEAUDOIN A., GRANGIER J.P., HUGUET G. *et al.*, "Calydial participe à un Living Lab pour le développement de la télémédicine", *European Research in Telemedicine*, vol. 3, pp. 47–49, 2014.

[CLE 17] CLEMENSEN J., ROTHMANN M.J., SMITH A.C. *et al.*, "Participatory design methods in telemedicine research", *Journal of Telemedicine and Telecare*, vol. 23, no. 9, pp. 780–785, 2017.

[COL 14] COLLOMB C., "Le monde du computationnel", *Intellectica*, vol. 61, pp. 313–324, 2014.

[CRA 04] CRAIG B.A.D., "Human feelings: why are some more aware than others?", *Trends in Cognitive Science*, vol. 8, pp. 239–241, 2004.

[DAR 13] D'ARRIPE A., ROUTIER C., "Au-delà de l'opposition entre savoir profane et savoir expert: une triangulation des méthodes", *Recherches qualitatives*, vol. 15, p. 230, 2013.

[DAS 95] DASTON L., "The moral economy of science", *Osiris*, vol. 10, pp. 3–24, 1995.

[DUB 14] DUBEY G., "Les nouvelles technologies en autonomie de santé: un déplacement des frontières de la connaissance", *Annales des mines–Réalités industrielles*, vol. 4, pp. 82–88, 2014.

[DUB 15] DUBEY G., "Le dessous des traces: pour une socio-anthropologie de l'e-santé", *Eä: Journal of Medical Humanities and Social Studies of Science*, vol. 7, nos 1–2, pp. 29–41, 2015.

[ELW 12] ELWYN G., FROSCH D., THOMSON R. *et al.*, "Shared decision making: a model for clinical practice", *Journal of General Internal Medicine*, vol. 27, pp. 1361–1367, 2012.

[ESC 14] ESCHALIER R., D'AGROSA BOITEUX M.C., MANNENQ P.H. *et al.*, "Cardiauvergne: service de télésurveillance et de coordination des soins des insuffisants cardiaques", *European Research in Telemedicine*, vol. 3, no. 4, pp. 169–176, 2014.

[FAS 09] FASSIN D., "Les économies morales revisitées", *Annales Histoire Sciences Sociales*, vol. 6, pp. 1237–1266, 2009.

[FES 16] FESSI H., RONCO P., "Quelques réflexions sur l'hémodialyse quotidienne à domicile en 2016: les conditions du succès", *Revue FNAIR, La voix des dialysés et des transplantés*, vol. 145, pp. 37–38, 2016.

[FRA 14] FRANC S., DAOUDI A., POCHAT A. *et al.*, "Évaluation des bénéfices apportés par le système Télédiab2 dans la prise en charge par les médecins généalistes de diabétiques de type 2 chroniquement déséquilibrés de la grande couronne parisienne", *European Research in Telemedicine*, vol. 3, no. 4, p. 189, 2014.

[GAL 14] GALINIER M., "Étude OSICAT: Optimisation de la surveillance ambulatoire des insuffisants cardiaques par télécardiologie", *European Research in Telemedicine*, vol. 3, no. 1, p. 45, 2014.

[HAR 88] HARAWAY D., "Situated knowledge: the science question in feminism and the privilege of partial perspective", *Feminist Studies*, vol. 14, no. 3, pp. 575–599, 1988.

[HES 02] HESSLOW G., "Conscious thought as simulation of behaviour and perception", *Trends in Cognitive Sciences*, vol. 6, pp. 242–247, 2002.

[HIM 09] HIMMA K.E., "Artificial agency, consciousness, and the criteria for moral agency: what properties must an artificial agent have to be a moral agent?", *Ethics and Information Technology*, vol. 11, no. 1, pp. 19–29, 2009.

[HOR 75] HORNE J., ÖSTBERG O., "A self-assessment questionnaire to determine morningness-eveningness in human circadian rhythms", *International Journal of Chronobiology*, vol. 4, no. 2, pp. 97–110, 1975.

[JAM 84] JAMES W., "What is an emotion?", *Mind*, vol. 9, pp. 188–205, 1884.

[JOH 06] JOHNSON D.G., "Computer systems: moral entities but not moral agents", *Ethics and Information Technology*, vol. 8, no. 4, pp. 195–204, 2006.

[KEL 15] KELL E., WAIT S., HARDING E. *et al.*, *A Policy Brief on Heart Failure in Europe*, The Health Policy Partnership, Brentwood, United Kingdom, 2015.

[KID 12] KIDHOLM K., EKELAND A.G., JENSEN L.K. *et al.*, "A model for assessment of telemedicine applications. MAST", *International Journal of Technological Assessment in Health Care*, vol. 28, no. 1, pp. 44–51, 2012.

[LAN 17] LANGARIZADEH M., MOGHBELI F., ALIABADI A., "Application of ethics for providing telemedicine services and information technology", *Medical Archives*, vol. 71, no. 5, pp. 351–355, 2017.

[LEH 04] LEHOUX P., "The use of technology at home: what patient manuals say and sell vs. what patients face and fear", *Sociology of Health and Illness*, vol. 26, no. 5, pp. 617–644, 2004.

[LEU 14] LEURENT P., BAILLY J.Y, BAHLOUL A. *et al.*, "La télédiabétologie: à l'ère de la télémédecine, qu'est-ce que la télédiabétologie?", *European Research in Telemedicine*, vol. 3, pp. 177–179, 2014.

[MIS 06] MISSONNIER S., "Périnatalité prénatale, incertitude et anticipation", *Adolescence*, vol. 1, pp. 207–224, 2006.

[MOO 06] MOOR J.H., "The nature, difficulty, and importance of machine ethics", *IEEE Intelligent Systems*, vol. 21, no. 4, pp. 18–21, 2006.

[MOR 12] MORGAN S., YODER L.H., "A concept analysis of person-centered care", *Journal of Holistic Nursing*, vol. 30, no. 1, pp. 6–15, 2012.

[OGD 15] OGDEN R., MOORE D., REDFERN L. *et al.*, "The effect of pain and the anticipation of pain on temporal perception: a role for attention and arousal", *Cognition and Emotion*, vol. 29, pp. 910–922, 2015.

[OUD 15] OUDSHOORN N., MATTHUIS I., "Technologies of compliance? Telecare technologies and self-management of COPD patients", *Eä Journal of Medical Humanities & Social Studies of Science and Technology*, vol. 7, no. 1, pp. 63–86, 2015.

[PAR 10] PARÉ G., MOGADEM K., PINEAU G. *et al.*, "Clinical effects of home telemonitoring in the context of diabetes, asthma, heart failure and hypertension: a systematic review", *Journal of Medical Internet Research*, vol. 12, no. 2, p. e21, 2010.

[PEN 12] PENFORNIS A., FRANC S., BENHAMOU P.Y. *et al.*, "P298 Étude Télédiab 1: l'utilisation effective de DIABÉO est-elle un facteur déterminant dans l'amélioration du contrôle glycémique?", *Diabetes & Metabolism*, vol. 38, no. 2, pp. A100–A101, 2012.

[PEZ 08] PEZZULO G., "Coordinating with the future: the anticipatory nature of representation", *Minds and Machines*, vol. 18, no. 2, pp. 179–225, 2008.

[POL 05] POLLATOS O., KIRSCH W., SCHANDRY R., "On the relationship between interoceptive awareness, emotional experience, and brain processes", *Cognitive Brain Research*, vol. 25, pp. 948–962, 2005.

[RYG 12] RYGH E., ARILD E., JOHNSEN E. *et al.*, "Choosing to live with home-dialysis-patients' experiences and potential for telemedicine support: a qualitative study", *BMC Nephrology*, vol. 13, no. 13, p. 3, 2012.

[SAC 96] SACKETT D.L., ROSENBERG W.M., GRAY J.A., HAYNES R.B. *et al.*, "Evidence based medicine: what it is and what it isn't", *BMJ*, vol. 312, no. 7023, pp. 71–72, 1996.

[SCH 14] SCHIRR-BONNANS S., TURNIN M., CHAUCHARD M. *et al.*, "EDUC@DOM: évaluation de l'efficacité d'un programme de télésurveillance au domicile et de son coût dans la prise en charge de patients diabétiques de type 2 (DT2) recrutés à partir d'un réseau de soins", *European Research in Telemedicine*, vol. 3, no. 1, p. 41, 2014.

[SOR 13] SORENSEN J.T., CLEMMENSEN P., SEJERSTEN M., "Telecardiology: past, present and future", *Revista Española de Cardiologia*, vol. 66, no. 3, pp. 212–218, 2013.

[SPE 14] SPEYER É., BRIANÇON S., JACQUELINET C. *et al.*, "Qualité de vie des personnes en insuffisance rénale chronique terminale en France en 2011", *Bulletin Épidémiologique Hebdomadaire*, vols 37–38, pp. 623–630, 2014.

[STH 06] ST-HILAIRE C., "Télésanté: lignes directrices cliniques et normes technologiques en téléréadaptation, Agence d'évaluation des technologies et des modes d'intervention de santé", *ETMIS*, 2006.

[SZP 10] SZPUNAR K., "Episodic future thought: an emerging concept", *Perspectives on Psychological Science*, vol. 5, no. 2, pp. 142–162, 2010.

[WAD 92] WADE J.E., SHERBOURNE C.D., "The MOS 36-item short-form health survey (SF-36)", *Medical Care*, vol. 30, pp. 473–483, 1992.

[WAL 17] WALKER C., HOWARD K., MORTON R.L., "Home hemodialysis: a comprehensive review of patient-centered and economic considerations", *ClinicoEconomics and Outcomes Research*, vol. 9, pp. 149–161, 2017.

[WHO 13] WHOLEY D.R., CHRISTIANSON J., WHITE K. *et al.*, "Improving chronic disease care by adding laypersons to the primary care team: a parallel randomized trial", *Annals of Internal Medicine*, vol. 159, no. 3, pp. 176–184, 2013.

## Reports and studies

[AGE 12] AGENCE NATIONALE D'APPUI À LA PERFORMANCE DES ÉTABLISSEMENTS DE SANTÉ ET MÉDICO-SOCIAUX (ANAP), La télémédecine en action: 25 projets passés à la loupe. Un éclairage pour le déploiement national. Tome 2: monographies, May 2012.

[BAR 14b] BARNSTEINER J., DISCH J., WALTON M.K., Person and family centered care, SIGMA Theta Tau International, Center for Nursing, 2014.

[BRI 18] BRITISH HEART FOUNDATION, CVD statistics – BHF UK factsheet, British Heart.

[CAH 13] CAHIERS I.P., Innovation prospective, no. 2, Le corps, nouvel objet connecté. Du *quantified self* à la m-santé: les nouveaux territoires de la mise en données du monde, Foreword, 2013.

[COU 15] COURT OF AUDITORS. Rapport sur l'application des lois de financement de la sécurité sociale, Chapitre X: l'insuffisance rénale chronique terminale: favoriser des prises en charge plus efficientes, Report, pp. 345–379, 2015.

[DGO 13] DGOS (DIRECTION GÉNÉRALE DE L'OFFRE DE SOINS), Principaux enseignements du bilan des PRT et du recensement des projets de télémédecine, DGOS, Report, 2013.

[DGO 15] DGOS (DIRECTION GÉNÉRALE DE L'OFFRE DE SOINS), Les chiffres clés de l'offre de soins, DGOS, Report, 2015.

[FRE 16] FRESENIUS MEDICAL CARE, International Strategic Development, 2016, available at: https://ercpa.eu/index.php/facts-figures/.

[GAR 98] GARROS B., Conférence nationale de santé 1998, rapport du HCSP: Contribution du HCSP aux réflexions sur la lutte contre l'iatrogénie, Report, no. 25, p. 9, December 1998.

[HAU 10] HAUTE AUTORITÉ DE SANTÉ (HAS), Les conditions de mise en œuvre de la télémédecine en unité de dialyse médicalisée, Synthèse et recommandations, January 2010.

[HAU 13] HAUTE AUTORITÉ DE SANTÉ (HAS), Rapport d'évaluation médico-économique. Efficience de la télémédecine: état des lieux de la littérature internationale et cadre d'évaluation, Report, July 2013.

[LIV 15] LIVRE BLANC DE LA SANTÉ CONNECTÉE DU CONSEIL NATIONAL DE L'ORDRE DES MÉDECINS, Santé connectée: de la e-santé à la santé connectée, White paper, 2015.

[OBS 12] OBSERVATOIRE PARITAIRE DES MÉTIERS DE L'INFORMATIQUE, DE L'INGÉNIERIE, DES ÉTUDES ET DU CONSEIL (OPIIEC), Étude sur les technologies de l'information au service des nouvelles organisations de soin, Création de valeur engendrée par le secteur des TIC santé en France, 2012.

[TEL 12] TELEMEDECINE 2020, Faire de la France un leader du secteur en plus forte croissance de la e-santé. Synthèse de l'étude sur les technologies de l'information au service des nouvelles organisations de soins, SYNTEC numérique, 2012.

[TEL 14] TELESCOPE RAPPORT, European code of practice for telehealth services Report, 2014.

[RAP 13] Rapport d'activité 2013 de l'ARS d'Auvergne, Report, 2013.

[SLO 06] SLOMAN A., Orthogonal recombinable competences acquired by altricial species, Technical report, University of Birmingham, 2006.

[SNI 14] SNITEM, Les innovations technologiques médicales. Dispositifs médicaux et progrès en diabète, pp. 6–7, November 2014.

[SYN 12] Synthèse du programme régional de télémédecine (PRT) de l'ARS Île-de-France, Diagnostic, pp. 11–13, 2012.

[WOR 86] WORLD HEALTH ORGANIZATION, Ottawa charter for health promotion, Ontario, 1986.

## Other written documents

[BUL 10] BULLETIN ÉPIDÉMIOLOGIQUE HEBDOMADAIRE, nos 42–43, pp. 425–440, November 2010.

[CAN 02] CANGUILHEM G., *Vie, Encyclopædia 16*, pp. 764–769, 2002, available at: http://www.universalis.fr/encyclopedie/vie/.

[CHR 13] CHRONOBIOLOGIE, LES 24 HEURES CHRONO DE L'ORGANISME, Dossier réalisé en collaboration avec le Dr Claude Gronfier, neurobiologiste à l'Institut Cellule Souche et Cerveau, Inserm U846, Lyon, December 2013, available at: www.inserm.fr.

[CRA 15] CRABBÉ J.-M., Neuropsychisme et rythmes biologiques, 2015, available at: www.sitemed.fr.

[KID 10] KIDHOLM K., BOWES A., DYREHAUGED S. *et al.*, The MAST manual, 2010.

[MCG 04] MCGRATH J.E., TSCHAN F., *Temporal Matters in Social Psychology: Examining the Role of Time in the Lives of Groups and Individuals*, American Psychological Association Publications, Washington, 2004.

[REI 14] Rein échos, La revue semestrielle gratuite de la ligue Rein et Santé, no. 15, pp. 1–60, October 2013–April 2014.

[THÉ 15] THÉBAUT J.-F., "L'actient", le patient, acteur de sa santé, available at: http://www.education-preventionsante.fr.

## Public interventions

[AUD 13] AUDRY A., Les conditions d'un déploiement industriel des solutions numériques pour la santé et la silver économie, Colloque Industries du numérique et de la santé autonomie, 2013.

[DAB 14] DABOT C., THOURET A., Télésurveillance en DP. À propos de la solution ap'Telecare de Tmm software, AG FEHAP Dialyse, Orléans, 2014.

[DUC 14] DUCHÊNE J., CERITD, Rencontres INRIA Industrie: Les sciences du numérique au service de la santé à domicile et de l'autonomie. Santé à domicile et autonomie : contexte, enjeux, perspectives, 2014, available at: www.ceritd.fr.

[FRA 15] FRANC S., LEPAPE G., Les nouveaux objets connectés des diabétiques, 2ème Forum du diabète, Sud Francilien, November 2015, available at: www.ceritd.fr.

[HIR 12] HIRSCH E., À nouveaux patients, nouveaux médecins, Éthique dans les usages numériques en santé, CNOM, Video, 2012.

[JAR 14] JARLAUD Y., Médias sociaux: le patient est mort, vive 'l'actient'!, Cinquièmes Assises des Technologies Numériques de Santé: "Numérique en santé: quelle place pour l'humain", 2014.

[JAR 15] JARLAUD Y., "Tenir compte de l'expérience patient", Troisièmes Assises de l'Innovation Thérapeutique, available at: www.hospitalia.fr/Rendez-vous-aux-3emes-Assises-de-l-Innovation-Therapeutique-Le-parcours-de-soins-a-l-heure-de-l-experience-patient_a712.html, 2015.

[MAR 14] MARCEAU J., Numérique en santé: quelle place pour l'humain?, 5èmes Assises des Technologies Numériques de Santé, Synthèse du colloque, p. 17, March 2014.

[WOY 16] WOYNAR S., Surveillance à domicile du patient insuffisant cardiaque: un accompagnement sur la durée et un confort au quotidien, Presentation, Castres, 2016.

## Webography

www.ama-assn.org
www.projetsante.com
www.eur-lex.europa.eu
www.lesmaisonsderetraite.fr
www.universalis.fr
www.coeur-recherche.fr
www.ercpa.eu
www.education-preventionsante.fr
www.numerique-sante.fr
www.bhf.org.uk/research/heart-statistics
www.ars.sante.fr
www.3hcoaching.com
www.hospitalia.fr

www.technologiesnumeriquessante.aromates.fr
www.ec.europa.eu/health/programme/policy/index_fr.htm
www.syntec.fr
www.recherche.telecom-bretagne.eu/gemsa/
www.gerontechnologies.com
www.renewinghealth.eu/en/
www.urc-eco.fr.
www.ticsante.com
www.invs.sante.fr
www.allodocteurs.fr
www.ameli-sante.fr
www.hospimedia.fr
www.tmm-groupe.com
www.catelparis.fr
www.cardiauvergne.com
www.afd.asso.fr
www.healthtechinsider.com
www.mdtmag.com
www.chu-mondor.aphp.fr
www.healthcare.orange.fr
www.sorin.com
www.inserm.fr
www.fnair.asso.fr
www.formaticsante.fr
www.topsante.com
www.ceritd.fr
www.education-preventionsante.fr
www.docplayer.fr
www.futura-sciences.com
www.lesporteurs.fr
www.universitedespatients.org
www.universite-esante.com
www.renaloo.com
www.tenon.aphp.fr/nephrologie-dialyse/
www.medtronic.com
www.has-sante.fr
www.ameli-sophia.fr
www.sante-medecine.commentcamarche.net
www.sitemed.fr
www.calydial.org
www.vivreavecundiabete.com

# Index

## H, I, J



## K, N



## O, P



## Q, R



## S



## T, V

Kolibri 360 GmbH
Ferdinand-Jühlke-Straße 7
99095 Erfurt, Deutschland
produktsicherheit@kolibri360.de